# Mad about
# Modern Physics

# Mad about Modern Physics

## Braintwisters, Paradoxes, and Curiosities

Franklin Potter

and

Christopher Jargodzki

John Wiley & Sons, Inc.

Published by John Wiley & Sons, Inc., Hoboken, New Jersey
Published simultaneously in Canada

Design and production by Navta Associates, Inc.

For general information about our other products and services, please contact our Customer Care Department within the United States at (800) 762-2974, outside the United States at (317) 572-3993 or fax (317) 572-4002.

Wiley also publishes its books in a variety of electronic formats. Some content that appears in print may not be available in electronic books. For more information about Wiley products, visit our web site at www.wiley.com.

*Library of Congress Cataloging-in-Publication Data:*

Potter, Frank, date
  Mad about modern physics : braintwisters, paradoxes and curiosities / Franklin Potter and Christopher Jargodzki.
    p. cm.
  Includes index.
  ISBN 978-1-63026-130-6
1. Physics--Popular works.  I. Jargodzki, Christopher II. Title
QC24.5.P68 2004
530—dc22

2004014941

10  9  8  7  6  5  4  3

To my late parents, who nourished my formative years and have now crossed that portal to another world.

F. P.


To my late grandmother—Zofia Lesinska, who instilled in me the idea that the visible world owes its being to the invisible one.

C. J.

# Contents

**Answers**

# Preface

This book of almost 250 puzzles begins where our first book, *Mad About Physics: Braintwisters, Paradoxes, and Curiosities* (2001) ended—with the physics of the late nineteenth and early twentieth centuries. The Michelson-Morley experiment of 1887, the challenges posed by atomic spectra and blackbody radiation, the unexpected discoveries of X-rays in 1895, radioactivity in 1896, and the electron in 1897 all loosened the protective belt of ad hoc hypotheses around the mechanistic physics the nineteenth century had so laboriously built. Anomalies and paradoxes abounded, ultimately necessitating a radical rethinking of the very foundations of physics and culminating in the theory of relativity and quantum mechanics. Numerous applications of these new and strange concepts followed very quickly as atomic and nuclear physics led to semiconductor devices on the small scale and nuclear energy on the large scale. Therefore we have developed a whole new set of challenges to tickle the minds of our scientifically literate readers, from science students to engineers to professionals in the sciences.

The challenges begin with the classical problem of getting a cooked egg into a bottle through a narrow bottleneck and back out again and progress gradually to the famous aging-twin paradox of the theory of special relativity and eventually reach problems dealing with the large-scale universe. In between, we explore the nature of time and of space as well as how the world of films and television tends to sacrifice physics for the sake of entertainment. We also consider some of the more startling questions in relativity. For example, we ask whether a person can go on a space journey out to a star 7,000 light-years distant and return while aging only 40 years! And we certainly want to emphasize the practical applications of microphysics through an examination of some properties of exotic fluids, unusual motors running on air or on random motion, as well as thermal, electrical, and photonic properties of materials in a challenging journey into the atomic world.

Particularly important microworld challenges include: What happened to Schrödinger's cat? Can a cup of coffee be the ultimate quantum computer? Why is a Bose-Einstein condensate a new state of matter? Why is quantum mechanical coherent scattering so important in developing new detectors for neutrinos and gravitational waves? When we reach the nucleus, there are challenges about the accuracy of carbon-14 dating, the reason for neutron decay, and the amount of human radioactivity. Then our journey reverses as we reach for the stars to consider Olbers' paradox about why the night sky is dark instead of bursting with light, how gravitational lensing by galaxies works, and what the total energy in the universe might be. This book finishes with a potpourri of challenges from all categories that ranges from using bicycle tracks in the mud to determine the direction of travel, to analyzing water-spouting alligators, and ending with a space-crawling mechanical invention that seems to defy the laws of physics.

The puzzles range in difficulty from simple questions (e.g., "Will an old mechanical watch run faster or slower when taken to the mountains?") to subtle problems requiring more analysis (e.g., "Is the Bragg scattering of X-rays from an ideal crystal a coherent scattering process?") Solutions and more than 300 references are provided, and they constitute about two-thirds of the book.

As these examples demonstrate, most of the puzzles contain an element of surprise. Indeed, one finds that commonsense conjecture and proper physical reasoning often clash throughout this volume. Einstein characterized common sense as the collection of prejudices acquired by age eighteen, and we agree: at least in science, common sense is to be refined and often transcended rather than venerated. Many of the challenges were devised to undermine physical preconceptions by employing paradoxes (from the Greek *para* and *doxos,* meaning "beyond belief") to create cognitive dissonance. Far from being simply amusing, paradoxes are uniquely effective in addressing specific deficiencies in understanding. Usually the contradiction between gut instinct and physical reasoning for some people will be so painful that they will go to great lengths to escape it even if it means having to learn some physics in the process.

Philosopher Ludwig Wittgenstein considered paradoxes to be an embodiment of disquietude, and as we have learned, these disquietudes often foreshadow revolutionary developments in our thinking about the natural world. The counterintuitive upheavals resulting from relativity theory and quantum mechanics in the twentieth century only enhanced the reputation of the paradox as an agent for change in our understanding of physical reality.

Such disquietudes, rather than unexplained experimental facts, writes Gerald Holton in *Thematic Origins of Scientific Thought,* were what led Einstein to rethink the foundations of physics in his three papers of 1905. Each begins with the statement of formal asymmetries of a predominantly aesthetic nature, then proposes a general postulate, not derivable directly from experience, that removes the asymmetries. For example, in the paper on the quantum theory of light, formal asymmetry existed between the discontinuous nature of particles and the continuous functions used to describe electromagnetic radiation. As Holton notes, "The discussion of the photoelectric effect, for which this paper is mostly remembered, occurs toward the end, in a little over two pages out of the total sixteen." Consistent with this approach is Einstein's statement in *Physics and Reality* (1936), "We now realize . . . how much in error are those theorists who believe that theory comes inductively from experience," and later in *The Evolution of Physics* (1938), coauthored with the Polish physicist Leopold Infeld, "Physical concepts are free creations of the human mind, and are not, however it may seem, uniquely determined by the external world."

As another sore point, the term "quantum *mechanics*" is really a misnomer: quantum systems cannot be regarded as made up of separate building blocks. In the helium atom, for instance, we do not have electron A and electron B but simply a two-electron pattern in which all separate identity is lost. This indivisible unity of the quantum world is paralleled by another kind of unity—between subject and object. Is light a wave or a particle? The answer seems to depend on the experimental setup. In the double-slit experiment, the observations of light yield characteristics of the box and its slits as much as of light itself. Is reality then observer-dependent? And would this justify Einstein's insistence on the power of pure thought in the construction of physical reality? Modern physics seems particularly adept at generating such disquietudes. If that's the case, then perhaps the word *Mad* in the title of our book should not be construed as a mere metaphor!

# Acknowledgments

We all "stand on the shoulders of giants" as we develop our minds to become individuals living today on our planet Earth. And we owe so much to so many people that we cannot acknowledge all of them.

Franklin Potter would like to express appreciation to his wife, Patricia, and their two sons, David and Steven, for their love and inspiration through many wonderful years of family adventures. He also treasures the numerous inspiring physics discussions over the decades with many friends and colleagues: Howard G. Preston, Gregory Endo, Fletcher Goldin, David M. Scott, John Priest, Lowell Wood, Julius S. Miller, George E. Miller, Leigh H. Palmer, Charles W. Peck, Myron Bander, Joseph Weber, Richard Feynman, Willard Libby, Edward Teller, and Kamal Das Gupta.

Christopher Jargodzki would like to express appreciation to Myron Bander of the University of California at Irvine; Stephen Reucroft of Northeastern University in Boston; and James H. Taylor of Central Missouri State University in Warrensburg. His interactions with close to twenty thousand students (and counting!) in his classes at UC Irvine, Northeastern University, and CMSU have been, over the years, never-ending sources of stimulation, as well as occasional exasperation. In fact, the present volume got its start in 1975 when one of us (C. J.), still a graduate student at UC Irvine, put together a proposal for a book of paradoxes in modern physics, partly to allay his own exasperation with the koanlike conundrums that abound in modern physics. Alas, the project had to wait several decades for the author to mature and join forces with Franklin Potter in our joint inquiry into the nature of physical reality. The authors hope that physical reality is duly impressed with their efforts.

Both authors sincerely thank Kate C. Bradford, senior editor at John Wiley & Sons, Inc., who continues to support our paradoxical adventures into the world of physics.

# To the Reader

These puzzles are meant to be fun. How many puzzles you solve is not as important as how many you enjoy thinking about. Some of them are even challenging to research physicists, and some were generated by research articles that have appeared only recently in physics journals, so these topics may not have been part of physics just 10 years ago! It would be a rare reader who could provide detailed solutions to all the puzzles. Indeed, sometimes you may need to think a bit to even understand the answer. If we included all the steps, this book would double its present size. We offer no apologies, but we do try to provide all the key steps to make each answer complete on its own. If you find the puzzles perplexing and intriguing, we have succeeded in our mission.

*Mad about Modern Physics* can be read with profit by anyone who has had some exposure to a year of introductory physics and is eager to learn more about its applications and its more recent discoveries. Most puzzles are nonmathematical in character and require only a qualitative application of fundamental physics principles. Many physics concepts are defined directly or indirectly in the questions or in the answers, so they can be found with the aid of the index. However, even someone who knows the subject will quickly realize that the application of physics to the real world can be quite challenging, and in this sense this is not an elementary book.

More than three hundred follow-up references provide further resources for interested readers. These references—to journal research papers, books, and magazine articles—are included with only some of the puzzles, typically those that are either controversial or that involve relatively new concepts. There was no space to include a more complete list of references. Consequently we had to make choices, and we apologize to the authors whose work may have been left out or inadvertently overlooked.

Any errors are solely those of the authors, and we would appreciate your communications via e-mail to Franklin Potter (see www.sciencegems.com) with regard to the puzzles and their answers.

# 1 | The Heat Is On

SCIENCE IN THE HOME CONTRIBUTES IMMENSELY to our everyday repertoire of activities, although most of us are unaware of exactly how science does so. Physics, in particular, is all around us and plays a crucial role in determining what we can and cannot do. One enjoyable activity for many people is cooking, which is an application of physics and chemistry to satisfy our gastronomical tastes. Or are physics and chemistry just other modes of cooking? We'll let you decide. Most of the challenges in this chapter involve physics from a high-school-level course. But be careful. Quick responses may be correct occasionally, but you should not rely on your intuition very much, for Nature, particularly in the kitchen, is nonintuitive for the most part. Anyone who has tried to make a soufflé can attest to how limited a recipe can be!

## I. Egg into a Bottle

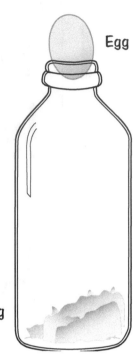

Egg

Perhaps the most intriguing physics-in-the-kitchen demonstration for all ages is getting a hard-boiled egg with the shell removed into a bottle that has an opening diameter smaller than the minimum diameter of the egg. One solution is to very carefully drop some bits of burning paper into the upright bottle and then place the egg at the opening. Soon, if the sequence is done with the correct tim-

Burning paper

ing, the egg will have the urge to go inside. What is the correct timing, and why does the egg have this urge?

## 2. Egg out of a Bottle

Perhaps the most *challenging* physics-in-the-kitchen demonstration for all ages is getting a hard-boiled egg with the shell removed out of a bottle that has an opening diameter smaller than the minimum diameter of the egg. Of course, one could cut up the egg with a knife inserted into the bottle and then pour out the pieces. However, we want the egg out whole and undamaged.

Long ago, physics professor Julius Sumner Miller, (who was Professor Wonderful on the early *Mickey Mouse Club* shows) was on the *Tonight Show* with host Johnny Carson and showed first how to get the egg into the bottle and then, taking no more than three seconds, had the same egg back in his hand. What is the procedure? (Hint: the same physics principles that put the egg into the bottle can get the egg out.)

## 3. Sugar

Add two cups of sugar to one cup of water in a saucepan and stir while heating slightly. All the sugar will dissolve. About how much total sugar will dissolve in one cup of water? What is the physics?

## 4. Kneading Bread

Bread made with yeast is usually kneaded—that is, drawn out and pressed together to create a distribution of the ingredients. Then the bread dough is set aside to "rise." Why is some bread then kneaded a second time and sometimes even a third time before baking?

## 5. Measuring Out Butter

Suppose you have a solid chunk of butter and a measuring cup in the kitchen. You desire to accurately measure one-half cup of butter chunks without melting them. What is a quick, easy way to do so? Often one encounters the statement in cookbooks that Archimedes' principle is being used. What is this principle, and why is the statement erroneous?

## 6. Milk and Cream

You are given two identical bottles, one with milk and the other with cream, both filled to the top. Quick now, which is heavier? And is light cream lighter than heavy cream?

Why is it that tea made with microwave-heated water doesn't taste as good as tea made with teakettle water? The main reason is that microwaves heat only the outer inch or so of the water all around the cup, because that's as far as they can penetrate. The water in the middle of the cup gets hot more slowly, through contact with the outer portions. When the outer portions of the water have reached boiling temperature and start to bubble, you can be tricked into thinking that all the water in the cup is that hot. But the average temperature may be much lower, and your tea will be shortchanged of good flavor.

—ROBERT L. WOLKE, *WHAT EINSTEIN TOLD HIS COOK: KITCHEN SCIENCE EXPLAINED*

If there were one drop of water less in the universe, the whole world would thirst.

—UGO BETTI, ITALIAN PLAYWRIGHT

## 7. Straw and Potato

A paper or plastic drinking straw can be pushed through an uncooked potato. Explain the physics. If you plan to try this demonstration, be sure that you take appropriate safety precautions—keep your hands and body out of harm's way.

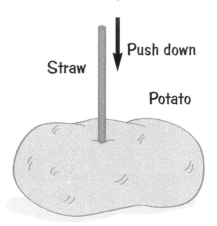

## 8. Blueberry Muffins

Marion loves to bake warm, fresh blueberry muffins, with the blueberries almost uniformly distributed throughout the muffin. She knows that if one simply prepares the batter and mixes in the blueberries, they may be uniformly distributed before entering the oven, but upon baking they will gravitate to lodge in the lower part of the muffin. How does she prevent this natural downward drift?

## 9. Can of Soup

Some people buy canned soup and store the cans in the cupboard. Some people even turn these soup cans upside down for storage. If we open a can of soup that was stored in the upright position by removing the top, quite often all the concentrated ingredients are on the bottom and must be scooped out with a spoon. Even then, not all the concentrate is removed. Suppose, instead, we turn the same can upside down and open the bottom. Upon turning the can over, the soup simply rushes out into the pot. Why so?

## 10. Salt and Sugar

Salts have been used for thousands of years to preserve meats, and sugar has been used to preserve fruits and berries. How do they work?

## 11. Defrosting Tray

In catalogs and cookware stores one can buy a "miracle" defrosting tray advertised as made of an "advanced, space-age super-conductive alloy" that "takes heat right out of the air." How does this defrosting tray work?

## 12. Ice Cream Delight

Most of us have made ice cream or seen ice cream being made. Milk, eggs, sugar, and flavorings are slowly chilled. Terri likes to make ice cream in a simpler and more efficient way. Practicing proper safety precautions, she pours liquid nitrogen directly into the ingredients in a metal bowl. About equal volumes of liquid nitrogen and the mixture are used for ice cream or sorbet, and she stirs while adding the coolant until the ice cream is nicely stiff. Why does this method produce absolutely marvelous ice cream, and what is the physics here?

## 13. Cooking a Roast

For many types of meat—beef, pork, lamb, etc.—one can buy a roast from the butcher with or without the bone inside. Suppose we have two beef roasts of the same

The boiling temperature of water decreases about 1.9°F for every 1,000 feet above sea level. So in Denver, the mile-high city, water will boil at 202°F—that is, at 94.4°C. Temperatures above 165°F are generally thought to be high enough to kill most germs, so there is no danger on this account until you get to about 25,000 feet.

On the average we get about 9 (food) calories (kcal) of energy from each gram of fat and 4 calories from each gram of protein or carbohydrate. To lose a pound (454 g) of fat, we have to cut the food intake by 3,500 calories. The discrepancy in numbers is due to the fact that body fat is only about 85 percent actual fat, the rest coming from connective tissue, blood vessels, and other things.

Light bounces off mirrors; microwaves bounce off metal. If what you put in the microwave oven reflects too many microwaves back instead of absorbing them, the magnetron tube that generates the microwaves can be damaged. There must always be something in the oven to absorb microwaves. That's why you should never run it empty.

Metals in microwave ovens can behave unpredictably. Microwaves set up electrical currents in metals, and if the metal object is too thin it may not be able to support the current and will turn red hot and melt. And if it has sharp points, it may even act like a lightning rod and concentrate so much microwave energy at the points that it will send off lightning-like sparks.

—ROBERT L. WOLKE, *WHAT EINSTEIN TOLD HIS COOK: KITCHEN SCIENCE EXPLAINED*

weight of 4.4 pounds (2 kg) and cook them in identical ovens at the same temperature. One roast has the bone in and the other does not. Which roast cooks faster? Why?

## 14. Cooking Chinese Style

Estimates of Chinese meals include more than 3,000 varieties, possibly more meal types than the total number of meals by all other cultures combined. Many of the Chinese dishes use meats cut into small cubes or other small volumes. Certainly, these small volumes are much easier to eat with chopsticks. Are there any significant scientific reasons for cutting up the meats into small volumes?

## 15. Baked Beans

If you buy dry beans in bulk, they must be soaked in water overnight in a covered container before they are ready to be baked. To bake them without soaking would require an enormous amount of cooking time. An alternative preparation is to "parboil" them in a cooking pot—that is, simmer them. Simmer means "to be on the verge of boiling."

How does one know that the beans have simmered enough? The test involves good physics. Take up a few beans in a spoon and, after making sure that no liquid is in the spoon, blow a stream of air gently with pursed lips against the beans. If the bean skin cracks, the beans are ready for baking. Why must the lips be pursed, and why do the bean skins then crack open?

## 16. Ice Water

Normally, to cool a pitcher of water quickly, one adds ice. The ice floats at the top. Suppose one could add the same amount of ice so it could be held in the water at the bottom of the pitcher. Which technique would lead to faster cooling of the water?

## 17. Peeling Vegetables

A friend of ours peels ripe tomatoes by impaling the tomato on a fork, then holding it over a gas flame and rotating gently. If you try this procedure, use appropriate safety procedures to protect your eyes and body.

Peeling fresh beets is also a messy chore. Their colored liquid stains everything, including your fingers. Another friend of ours peels fresh beets by first boiling them, then immediately holding them under cold water with a fork. What is the physics in both of these methods used for preparing vegetables for peeling?

## 18. Igniting a Sugar Cube

Sugar burns in air. But igniting a sugar cube is much more difficult than expected. Put a sugar cube on the end of a toothpick and bring a lighted match flame under a remote corner. The sugar melts instead of burning, and the brown, gooey stuff is caramel.

However, we wish to burn the sugar, not melt it! We want to see it on fire with a flame of its own. Why is this process so difficult to achieve? How can we succeed in lighting the sugar cube with the burning match?

## 19. Water Boiling

An open pot of water is boiling on the kitchen stove. Sprinkle some room-temperature table salt (which

A standard 12-ounce aluminum can, whose wall surfaces are thinner than two pages from this book (about 0.00762 cm), withstands more than 90 pounds of pressure per square inch—three times the pressure in an automobile tire.

—WILLIAM HOSFORD AND JOHN DUNCAN, "THE ALUMINUM BEVERAGE CAN," *SCIENTIFIC AMERICAN*, SEPTEMBER 1994

Decaffeinated coffee still contains caffeine! A regular cup of coffee has 80 to 135 milligrams of caffeine. For a coffee to be considered decaffeinated, at least 97 percent of the coffee's caffeine must be removed. Testing shows that decafs typically have 2 to 6 milligrams of caffeine per cup.

An object at room temperature (20°C) emits radiation with a peak at the wavelength 9.89 micrometers, roughly .01 mm, in the infrared region of the electromagnetic spectrum.

For an isolated water molecule the H-O-H angle is 104.5°. In ice each water molecule forms hydrogen bonds to four nearest neighbors in a tetrahedral arrangement. The tetrahedral bond geometry explains the openness and relatively low density of ice (i.e., why water expands upon freezing). In ice the H-O-H angles are nearly the same as the perfect tetrahedral angle of 109.5°.

contains mostly NaCl and some KCl) into the clear boiling water, and the boiling ceases. Isn't it amazing how the water ceases its boiling as the salt warms up! Can you explain the physics? What is the surprise here?

## 20. Put the Kettle On

Bring water to a boil in a teakettle with a spout. Let it cook! Now watch the mouth of the spout carefully. What do you see? Can you see the water vapor come out?

## 21. The Watched Pot

You have probably heard the expression "A watched pot never boils." Is this statement correct physics? That is, when would this statement be good physics? (Hint: One should interpret the phrase "never boils" here to mean that the cooking takes a longer time.)

## 22. Ice in a Microwave

The microwave oven emits microwaves that are absorbed by water molecules in food. Microwaves make the polar water molecules rotate or oscillate, and their "friction" within the material converts some of this kinetic energy into thermal energy to raise the temperature of the food.

Suppose you made an ice block that had liquid water trapped in a large cavity inside and then you placed the block into a microwave oven. Could the trapped water be brought to a boil while the ice remained ice?

## 23. The Glycemic Index

The glycemic index is an important number for anyone concerned with the conversion of food to blood sugar (sucrose), for the gylcemic index gives the measured rate of this conversion process. The higher the glycemic index value, the faster the conversion rate to sucrose. There are types of sugar molecules other than sucrose. Glucose, for example, is normally the standard reference for the conversion rate to sucrose, with a value of 100.

Some sample values of the glycemic index for foods are: brown rice, 59; white rice, 88; table sugar, 65; grapefruit, 25; spaghetti, 25 to 45; potato, boiled, 55; potato, baked, 85; and dates, 103. Brown rice has more outer layer intact than white rice, so its lower value is evident. But why would a baked potato have a much higher glycemic index than a boiled potato? And how could the value for dates, or any food, be higher than 100?

## 24. Electric Pickle

Some specialty and novelty stores sell an electrical "appliance" that cooks hot dogs between two metal electrodes. A protective cover with a safety interlock

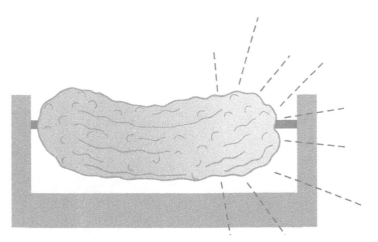

Night cooling by evaporation of water and heat radiation had been perfected by the peoples of Egypt and India, and several ancient cultures had partially investigated the ability of salts to lower the freezing temperature of water. Both the ancient Greeks and Romans had figured out that previously boiled water will cool more rapidly than unboiled water, but they did not know why; boiling rids the water of carbon dioxide and other gases that otherwise retard the lowering of water temperature.

—TOM SHACHTMAN, *ABSOLUTE ZERO AND THE CONQUEST OF COLD*

Interestingly, microwave ovens are not very good at melting ice. The water molecules in ice are bound pretty tightly together into a crystal lattice, so they can't flip back and forth under the influence of microwaves' oscillation.

Although it flies in the face of common sense, people with more insulation—fat—whose body core is better protected from the cold, may feel cold more quickly than thinner people with less protection. The reason is that insulation keeps heat in the core, away from the skin, which gets cold. When the skin gets cold, you feel cold. Paradoxically, women may feel colder than men because women are better insulated.

—JAMES GORMAN, "BEYOND BRR: THE ELUSIVE SCIENCE OF COLD," THE NEW YORK TIMES, FEBRUARY 10, 2004

The main compartment of a refrigerator should always be below 40°F (4.4°C). Above that temperature, bacteria can multiply fast enough to be dangerous.

closes over the device before electrical energy in the form of a standard AC current can be applied. Suppose that instead of a hot dog one places a pickle between the electrodes. When the room lights are dimmed, the pickle glows impressively, predominantly at one end. What is the physics, and what might the glow look like?

## 25. Space-Age Cooking

Microwave ovens were probably the first new method for making heat for cooking in more than a million years. In addition, two newer methods have become available for the kitchen. Magnetic induction cooktops have been available for about fifteen years in Europe and Japan and are now becoming known in the United States. And for the modern chef, cooking with light in a "light oven" has been done since the mid-1990s and may become a fad in the immediate future. How do both of these cooking sources work?

# 2 | Does Anybody Really Know What Time It Is?

**W**HAT IS TIME?" ST. AUGUSTINE FAMOUSLY wrote. "If no one asks me, I know. But if I wanted to explain it to one who asks me, I plainly do not know." Time itself is a strange quantity to some people. To many of us, time never seems to be going at the right rate— sometimes too fast, sometimes too slow. In some parts of the world, promptness and being on time are important aspects of the local culture. In other regions, time is almost irrelevant. In this chapter, we have created a mixture of familiar challenges and many new ones in preparation for later chapters in which time shares its role with space as a major ingredient of motion, chapters that look at concepts such as the space-time of the special theory of relativity and the world of astrophysics.

## 26. January Summer

Contrary to the popular belief that Earth is closest to the Sun on about June 23 or possibly December 22 each year, the date of perihelion actually falls between January 2 and January 5! In the Northern Hemisphere, we experience winter on this January date because the North Polar axis is tilted away from the Sun. The Southern Hemisphere enjoys a warm summer at this time. Will the Northern Hemisphere ever enjoy summer in January?

## 27. Proximity of Winter Solstice and Perihelion

Earth reaches perihelion—the point in its orbit when it's closest to the Sun—between January 2 and 5, depending on the year. That's about two weeks after the December solstice, December 21 or 22. Thus winter begins in the Northern Hemisphere at about the time that the Earth is nearest the Sun. Is there a reason why the times of solstice and perihelion are so close, or is this a coincidence?

## 28. Earth's Speed

The time interval required for Earth to travel from the autumnal equinox to the vernal equinox (approximately 179 days) is less than the time interval from the vernal to the autumnal equinox (roughly 186 days). Why?

## 29. The Equinox Displaced

At the time of the spring equinox (usually March 20) or the fall equinox (September 22 or 23), night and day are supposed to be of equal duration. But according to the almanacs of sunrise and sunset times, on the dates of the equinoxes, daytime is longer by 8 to 10 minutes. How come?

## 30. The Dark Days of December

At latitude 40 degrees north, earliest sunset occurs on about December 8 and latest sunrise on about January 5. The shortest day of the year, the winter solstice, is December 21 or 22. Why are all these dates not the same?

## 31. Days of the Year

The length of the year (i.e., the interval of time between two successive passages of Earth through the same point in its orbit) is about 365.2422 days. How many entire rotations on its own axis does Earth execute during that time?

## 32. Leap Years

Every four years, in years divisible by four, is a leap year, when an extra day is added to the month of February, except years divisible by 100. For example, 1700, 1800, and 1900 were not leap years, yet 2000 was a leap year. Why?

## 33. Full Moons

Is the interval of time between one full Moon and the next equal to 28 days?

## 34. Moon Time

Cheryl is sitting at a desk in an office and the clock shows 12:20 and the Moon is seen through the window as a thin crescent with the open side pointing downward to the right. What do you make of this scene? Where could the Sun be?

The minute first appeared as a division of the hour about A.D. 1320 in Paris editions of the so-called Alfonsine Mean Motion Tables, sponsored by King Alfonso the Wise of Spain. But the *idea* of the minute was implicit all the time in a method of reckoning used by early astronomers. They employed a system of sexagesimal fractions, first devised by the Babylonians, based on successive powers of 60. Any unit could be divided into 60 parts; these were called in Latin "partes minutae primae," or "first very small parts," yielding the word "minute". A minute in turn was eventually divided into 60 "partes minutae secundae," hence the word "second."

—Q & A, SCIENCE TIMES,
*THE NEW YORK TIMES*,
DECEMBER 13, 1983

When it comes to procrastinating, I do it right away!

—ANONYMOUS

In Wicca, February 2 (Groundhog Day) is one of the four "greater sabbats" that divide the year at the midpoints between the solstices and equinoxes.

Sundials tell Sun time while clocks tell mean time. The true Sun leads or lags the mean Sun, crossing the meridian from 16 minutes, 25 seconds earlier than the mean Sun (in early November) to 14 minutes, 20 seconds later (in February). Only on or about April 16, June 14, September 2, and December 25 are the true and mean Suns together as they cross the meridian.

The angle between the Equator and the ecliptic (i.e., the plane of Earth's orbit), also known as the tilt of the globe, was 23° 26' 32" in 2002. Through the ages, this value varies between 21° and 28°. At present it goes down by 0.47" per year.

## 35. Lunar Calendar

Although there have been numerous calendars over the millennia of civilizations, they fall into two basic types, solar and lunar calendars. Today, while practically everyone uses the solar calendar with 365.2422 days per tropical year, rice farmers in many parts of the world continue to use the lunar calendar based on a 29.53-day lunar month. Can you figure out a scientific reason why?

## 36. The Sandglass

For a sandglass timer one could simply have a straight glass or plastic tube with equally spaced markings and then the whole tube would be inverted to start the time measurement. Why do ruled sandglasses have a tapered "hourglass" shape instead?

## 37. Old Watch

Lenni has an old mechanical watch in pristine condition that has an internal balance wheel that operates perfectly. She takes a drive into the mountains. Will the watch run fast or slow?

## 38. Reading a Digital Timer

Many digital timers show the elapsed time to one-hundredth of a second. What is the minimum uncertainty in the value? What value should be reported?

## 39. Eternal Clocks?

There are laser and atomic clocks in special laboratory environments that are accurate to one second in 300 million years! Yet their lifetimes are typically less than 30 years. Some wristwatches run longer! There are mechanical clocks in development that could last about 10,000 years! But they would need periodic winding. Why do the laser and atomic clocks have such short lifetimes? How might one build a mechanical clock that would survive so long?

## 40. Room Light

Suppose there is a photodetector with a flash lamp at the exact center of a 3 m × 3 m × 3 m dark, barren room with reflective walls. The flash lamp flashes for one nanosecond. For simplicity, assume that the light is emitted isotropically in all directions when the lamp flashes. If the photodetector simply sums the light from all directions, what is its recorded intensity versus time? If the photodetector is an array capable of discerning different angular directions, what is the intensity versus time for several different directions? Suppose the lamp flashes for one microsecond. What now?

## 41. Right to Left Driving Switch

Suppose you live in a country in which the driving is on the right and there is to be a change to driving on the left. If highways with on-ramps and off-ramps,

More people are born on October 5 in the United States than on any other day. Not so surprising, as conception would have fallen on New Year's Eve.

If 23 students are in a classroom and you pick two at random, the probability that their birthdays (month and day) match is about 1/365. The probability that at least two of the 23 have the same birth date, however, is a trifle better than ½. The reason is that now there are 1 + 2 + 3 + . . . + 22 = 253 possible matching pairs.

—Martin Gardner, "Mathematical Games," *Scientific American* (October 1972)

There are 365 days in the year. Note the following:

$$365 = 10^2 + 11^2 + 12^2$$
$$= 13^2 + 14^2$$

Coincidence? Preestablished harmony? You be the judge!

and so on are built for driving on the right, will they work equally well for driving on the left? Of course, we must assume the same patterns of driving speeds as before.

## 42. Light Clock

Some museums and laboratories have a light clock with two parallel mirrors and a pulse of light bouncing back and forth repeatedly, retracing the same path over and over, keeping very accurate time as each complete transit of the light pulse is detected and counted. The mirror separation is usually about a meter or less, so a

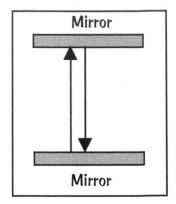

very large number of reflections occur during each second of time. Suppose this light clock is moved sideways parallel to the mirrors at a constant velocity, and assume that the light will continue to reflect off both mirrors during this sideward movement. Will the clock continue to keep accurate time?

## 43. Time Reversal

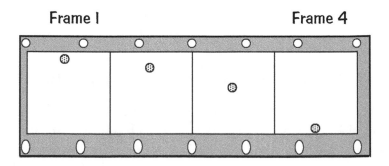

A movie is made showing successive frames for an object accelerating downward. If the sequence is run backward, the object accelerates (a) upward or (b) downward. Explain.

## 44. Molecular Clock

Different species of organisms have enormous regions of DNA that are the same or very similar. Humans and chimpanzees, for example, share about 98 percent of their DNA. We share much less of our DNA with rodents and amphibians and insects.

In a general way, the percentage of shared DNA might be a means to establish a molecular clock—that is, the more DNA that is shared, the more recent was the separation of the family tree. And, if by accident, the changes in the DNA happened to proceed at a common rate, then one could set up a timeline also.

However, the genetic changes do not occur with any regularity. Why not?

## 45. SAD

Most animals experience dramatic seasonal cycles: they migrate, hibernate, mate, and molt at specific times of the year. These cycles appear to be hardwired; they occur even when the temperature is held constant and the light and dark periods are varied. But humans are among the least seasonally sensitive creatures, having only a vestige of seasonal effects known as seasonal affective disorder (SAD), an extremely mild version of the cyclical responses animals experience. Only about 5 percent of adults overtly sense the seasonal changes and suffer from SAD during the winter days of longer darkness. Amazingly, light therapy—looking into a light that mimics sunlight—or merely sleeping until dawn helps

In a *Proustian moment* an unexpected smell or taste or perhaps a song from your past can unleash in you a raging torrent of realistic and graphic memory. The phrase recalls a scene in Marcel Proust's *Remembrance of Things Past* when a madeleine cake (a small, rich cookie-like pastry) enables the narrator to experience the past completely as a simultaneous part of his present existence: "And suddenly the memory revealed itself: The taste was that of the little piece of madeleine which on Sunday mornings at Combray (because on those mornings I did not go out before mass), when I went to say good morning to her in her bedroom, my aunt Leonie used to give me, dipping it first in her own cup of tea or tisane."

HOW TO FIND NORTH
USING A WATCH
In the Northern Hemi-
sphere, hold the watch
horizontal and point the
hour hand at the Sun.
Bisect the angle
between the hour hand
and the 12 o'clock mark
to get the north-south
line. If your watch is set
on daylight saving time,
use the midway point
between the hour hand
and one o'clock. The
farther you are from
the Equator, the more
accurate this method
will be.

the people with SAD in northern latitudes. Would these therapies be effective on people living at the Equator?

## 46. Two Metronomes

Suppose the timekeeping abilities of two identical metronomes are compared over several hours. They will drift faster or slower at different rates. When both metronomes are placed on a skateboard that moves freely horizontally, their drifts change gradually as they tend to synchronize. Each metronome has been subjected to the driving force of the other, the result being the phenomenon called "phase-locking" or "mode-locking." Suppose now that each metronome on the skateboard begins with different initial conditions, but one of the two metronomes is driven by perturbations that fluctuate *randomly* in time. Can the metronomes become synchronized?

## 47. Time Symmetry

The fundamental equations of physics—at least those that derive from symmetries in nature—all exhibit time symmetry because they are second-order differential equations. Newton's second law and Maxwell's equations are immediate examples. However, one can consider time running forward or backward. Even general-relativity equations formulated in tensor mathematics exhibit time symmetry. Assuming that all these equations are correct, must nature at its most fundamental level obey time symmetry? (Note: Entropy relations are not derived from a fundamental symmetry and therefore are excluded.)

# 3 | Crazy Circles

SPACE IS RELATED TO POSITION, DISTANCE, AND size and has its own paradoxes and influences. We live in a space of three dimensions, but our ability to visualize three-dimensional relationships among objects is not as easy as judging distance. Our brain activity relies on neural connections in a 3-D biomass that would probably become moronic if limited to two dimensions. However, robots usually operate in our 3-D space by following computer programs that maneuver in multidimensional configuration spaces that often far exceed three dimensions. Recent theoretical research in quantum physics hints that the natural world may be as large as 11-dimensional, with seven dimensions curled up too small for our senses, leaving the four dimensions of space-time. In this chapter we have created a mixture of familiar challenges and many new ones regarding space in preparation for a later chapter on the space-time of the special theory of relativity.

## 48. Spider and Fly

On a plane the shortest distance between two points is a straight line. Suppose a spider sits on a cube and wants to catch a fly sitting on the opposite face. How would you determine the path of shortest distance for the spider to crawl on the surface to catch the fly?

## 49. Moon Distance

In measuring the length of a 1-meter table with a meterstick to within 0.1 millimeter, the uncertainty in the measurement is one part in ten thousand. Metersticks, however, are usually inconvenient for measuring the distance to the Moon. Instead, a laser light pulse can be reflected from a stationary corner reflector on the Moon similar to the reflectors on bicycles, and the total duration of the pulse from Earth to Moon and back to Earth again is timed. What do you estimate for the uncertainty in the measurement for the Moon's distance? Which determination would you expect to have the greater distance uncertainty, the table length or the distance to the Moon?

## 50. Ideal Billiards Table

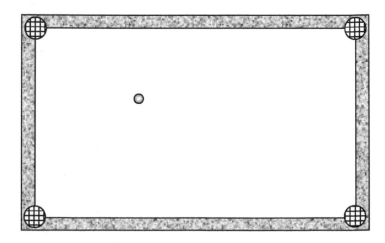

Suppose you have an ideal rectangular billiards table on which a ball collides with any wall (called the cushion) so that the angles of incidence and reflection are equal. Let there be pockets at the corners only. Describe how to shoot a given ball into a specific corner pocket with either zero, one, two, or three banks of the ball.

# 51. Wallpaper Geometry

Some of the old video games used an interesting but simple visual technique to extend the playing field. A character running off the right side of the screen then entered the left side while the background scenery remained fixed. That is, the right side edge is matched to the left side edge, and the top and bottom are matched also. One could even have a rectangular array of video screens, each right edge matched to a left edge, etc., each screen showing the same image. Faster systems later came along, and the scenery moved instead, and these 2-D views were eventually replaced by 3-D views.

Consider now a 3-D regular array of cubes touching face to face and top to bottom, the 3-D space analog to the old style 2-D video game. Let opposite cube faces be matched and imagine that these face surfaces are invisible. You are standing in one cube inside this space and look to your right. Behold! You see yourself!

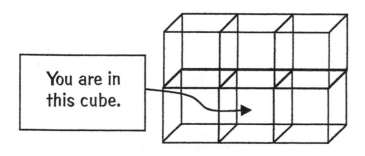

You are in this cube.

What exactly do you see? What do you see when looking upward?

## 52. Space-Filling Geometry

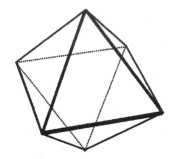
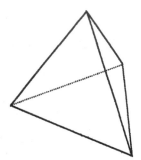

Cubes can be placed next to each other in three directions to fill all of 3-D space. Regular octahedrons can
fill 3-D space also. Spheres of the same radius cannot.
Can regular tetrahedrons fill all of 3-D space and leave
no gaps? Can regular dodecahedrons and regular icosa-
hedrons?

## 53. Archimedes' Gravestone

Archimedes' gravestone is said to have a sphere inside
a cylinder etched into the stone as well as the symbol π.
How are the two 3-D objects related if they have the
same radius? And why are they on his gravestone?

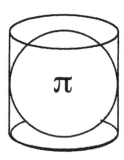

## 54. Brain Connections

The human brain has more than 100 billion neurons, with each neuron receiving input signals from 10 to 1,000 other neurons. Schematic representations of these connections in the brain always show an incredible web of lines representing the neurons, either as a 2-D or a 3-D image. Suppose you created a scaled-down computer model of this human brain using only 1 million neurons in a 3-D space. On average, how many input connections would each neuron have? What is the surprise here?

## 55. Configuration Space

Suppose we have a robotic arm that mimics the movements of a person's arm. The arm exists in the familiar 3-D physical space. Consider a simplification of the robotic arm that assumes just three connected parts: upper arm, forearm, and hand, all in the shape of straight rods that are connected. The body of the robot, including the shoulder, remains fixed in position. We wish to have the robotic arm touch a particular point-like object in the room. How many numbers are required in a computer program to describe the arm position?

## 56. Farmer Chasing a Goose

Farmers know that to catch a stray goose one does not run after the goose in an open field. A better strategy is to corner the goose. However, suppose the farmer and the goose are in an open field and they both run with the same speed, $V$, to provide us with some semblance of fair play. Furthermore, restrict the farmer to chasing the goose along the instantaneous line of sight to the goose. When will the farmer catch the goose?

THE LATE APPEARANCE IN ENGLISH OF THE WORD "SCIENTIST"
In 1840 William Whewell noted that there was no simple and natural way to refer to "a cultivator of science in general." He was, he concluded, inclined to call him "a scientist." Before Whewell scientists tended to refer to each other as philosophers, or more fully, as natural philosophers. For this reason Newton's treatise on mathematical physics was given the title *The Mathematical Principles of Natural Philosophy* (1687).

The Indian mathematician Srinivasa Ramanujan (1887–1920) discovered an approximation to $\pi$ that is remarkable for its precision and conciseness: $(2143/22)^{1/4}$ = 3.14159265258 ... (to be compared with $\pi$ = 3.14159265358 ...).

DIVINE MADNESS
The word "theory" comes from the Greek word *theoria,* meaning "ecstatic contemplation of the truth," as exemplified in Plato's belief that "the greatest truths are those that come to us through divine madness."

# 57. A Spooky Refrigerator

Christina notices that food is disappearing from her refrigerator and yet her surveillance camera shows that no one is opening the door. Suppose our 3-D spatial world were really a 4-D spatial world, but we did not know anything about the existence of the fourth spatial dimension. There is still the single time dimension. Could a 4-D being remove food from her 3-D refrigerator without opening the refrigerator door?

# 58. Fractional Dimensions?

A point has zero dimensions. A line has one dimension. A plane has two dimensions. Space has three dimensions. Can something have 1.585 . . . spatial dimensions?

# 59. Platonic Solids

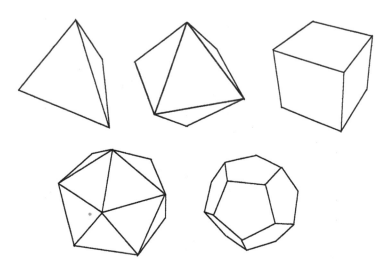

There are five 3-D regular polyhedrons called the Platonic solids: the regular tetrahedron (4 faces), the regular hexahedron (cube), the regular octahedron (8 faces), the regular dodecahedron (12 faces), and the regular icosahedron (20 faces). All these solids have a twofold rotational symmetry axis through the center of each edge—that is, a rotation about this axis by 180 degrees leaves the object looking the same as the initial view. But the regular tetrahedron does not have inversion symmetry.

If we intersect two identical regular tetrahedrons so their centers coincide, can the composite object have a twofold rotational symmetry axis? Can it have inversion symmetry?

# 60. Intersecting Spheres

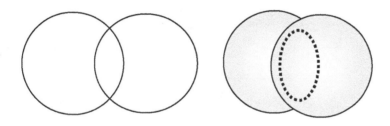

If in 2-D we intersect two circles (called one-spheres by mathematicians), the intersection is either a point, two points, or a circle. In 3-D the intersection of two spheres (each called a two-sphere) will be either a point, a circle, or a sphere. What can the intersection of two three-spheres be? And three three-spheres?

# 61. Arm Contortions

Normally, the rotation of an object about a fixed axis by 360 degrees brings the object back to its initial orientation. However, Barbara has the agility to do the following double rotation. She places a small object or book in her right hand, holding the object horizontal

The size of the Moon compared to the Earth is 3:11 (with accuracy of 99.9 percent). This Earth–Moon proportion is also precisely invoked by our two planetary neighbors, Venus and Mars. The closest : farthest distance ratio that each experiences of the other is, incredibly, 3:11 (with accuracy of 99.9 percent). Quite by chance, 3:11 is 27.3 percent, and the Moon orbits the Earth every 27.3 days, also the average rotation period of a sunspot.

—JOHN MARTINEAU, *A LITTLE BOOK OF COINCIDENCE*

The Roman numeral representing "five," symbolized by the letter V, derives from the shape of the space between the open thumb and index finger. The Roman numeral for "ten," the letter X, is actually two V's.

The biblical approximation of π is given in I Kings 7:23 and is repeated in 2 Chronicles 4:2. Both verses speak of a circular "sea of cast bronze" with a diameter of 10 cubits and a circumference of 30. The Greeks used a more accurate value of 22/7 (error 0.04 percent), and the Egyptians used a ratio of two squares 256/81 (error 0.6 percent).

The other day, I was walking my dog around my building . . . on the ledge. Some people are afraid of heights. Not me, I am afraid of widths.

—STEVEN WRIGHT, COMEDIAN

R. G. Duggleby, a biochemist at the University of Ottawa, found that the sum of π to the fourth power (97.40909 . . .) and π to the fifth power (306.01968 . . .) is e (i.e., 2.7182818 . . .) to the sixth power (403.42879 . . .), correct to four decimal places!

and noting its orientation in the room. While imagining a vertical axis from floor to ceiling, the book is moved inward first and then under the upper arm, keeping the book horizontal and rotating the object completely around this vertical axis back to its initial position. Her arm is now twisted. Can she untwist by rotating her arm a second time in the same direction?

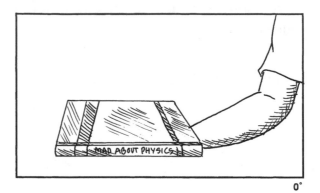

0°

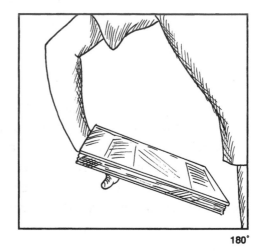

180°

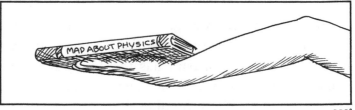

360°

## 62. The Rotating Cup

Place a cup with a handle on a shelf at eye height. Now walk in a straight line at a nearly constant speed past the cup, all the while rotating your head to observe the orientation of the cup. Notice what you see. The cup appears to rotate in the direction opposite your walking direction, at first very slowly, then quickly, then slowly again. Now consider yourself to be stationary and imagine the cup itself moving past in a straight line with constant speed. You could try to demonstrate this with the cup in your hand. What do you see now?

## 63. Space and Time Together

To explain Einstein's 1905 special theory of relativity and Minkowski's 1908 unification that combines three space dimensions and one time dimension into a four-dimensional space-time continuum, most introductory physics textbooks use a four-dimensional coordinate system, with three real coordinates for space and one imaginary coordinate for the time coordinate. Why not four real coordinates? Why not have three imaginary space coordinates and one real-time coodinate?

## 64. Space > 3-D?

Can you provide arguments for why space has three dimensions? Hint: Are planetary orbits stable in a space of $n$ dimensions, where $n > 3$? Is the hydrogen atom stable when $n > 3$?

The ancient composers of Vedic literature in India had to develop a method of evaluating square roots. The technique apparently evolved from a need to double the size of a square altar. One needs a square whose sides are the square root of 2. In the *Sulbasutras*, a collection that dictates the shapes and areas of altars and the location of the sacred fires, the square root of 2 is stated as 1.414215 . . . , an amazingly accurate value! The *Sulbasutras* were written between 800 and 500 B.C., making them at least as old as the earliest Greek mathematics. The Greeks, however, had no positional notation system. Hence their approximations of the square roots were rather crude.

—GEORGE GHEVERGHESE JOSEPH, *THE CREST OF THE PEACOCK: NON-EUROPEAN ROOTS OF MATHEMATICS*

# 4 | Fly Me to the Moon

**W**E LIVE IN A WORLD OBEYING THE RULES OF nature. But this natural world described by physics and the other sciences can be superseded and replaced by the imagination of the human mind. The artificial worlds created in the many forms of literature and in audio and visual renderings today cast a powerful influence on the minds of everyone in the modern world. In fact, more people prefer to live in these artificial fantasy worlds than in the real world than are willing to admit. In these challenges we focus on some of the "fuzzy science" prevalent in movies and television shows. In certain ways, the awareness of the correct science can enhance your enjoyment of the entertainment product, just like knowing how a bee communicates to the other bees in its hive enhances the beauty of the bee itself.

Jonathan Swift's
*Gulliver's Travels*,
published in 1726,
describes Gulliver's
many adventures,
including his "Voyage to
Laputa." Gulliver learns
that the scientists there
discovered two moons
of Mars, which revolve
around the planet at
distances from its cen-
ter equal to 3 and 5
Martian diameters.
When the moons of
Mars were discovered
by Asaph Hall in 1877, it
turned out that Swift
not only had the number
of the moons right, but
he also placed them
close to the actual
distances: 1.4 and 3.5
diameters of Mars.
The two moons, named
Phobos and Deimos, are
tiny. Phobos, which
measures 27 by 19 kilo-
meters, is shaped rather
like a potato. Deimos,
too, is oddly shaped,
and measures 15 by 11
kilometers.

# 65. Gunfight

Some TV programs and films have high drama scenes
based on a victim being shot by the pursuer and being
"blown backward" a meter or two by the projectile
impact. Is this dramatic response Hollywood hype, or
is there good physics here?

# 66. Body Cushion

A fall from a height of several stories onto pavement or
even onto a lawn will produce serious injuries or even
death. Yet we have seen the movie hero going over the
edge of a roof holding another human body in position
just beneath to cushion the fall on impact. Certainly,
collision with this second body is better than direct collision with the ground. What do you think about the
advantages here?

# 67. Cartoon Free Fall

So many of us in our youth learned the laws of nature
from cartoons. Some of us are still learning from
cartoons! The cartoon character steps forward off
a cliff and remains there in
suspension *until realizing the
situation,* then the acceleration
downward begins. As you
recall the scene, what violations
of physics can you discern?

# 68. Silhouette of Passage

When a cartoon character
smashes through a solid wall or
other object, we see the perfo-
ration as the crisp outline of the

character. What would a condensed matter physicist say about this cookie cutter type of material response?

## 69. Artificial Gravity

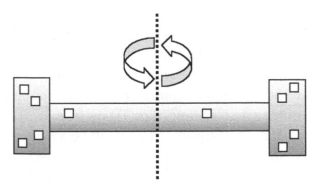

We all know that a body will tend to "float" around in a space station orbiting Earth or in a spaceship cruising at a constant velocity with respect to the stars. Some films depict a dumbbell-shaped space station rotating about an axis through its middle perpendicular to the long axis in order to provide artificial gravity. What interesting behavior patterns might be experienced by an astronaut who walks across the axis from one end to the other?

## 70. Small Wings

Space heroes who visit other planets have encountered alien beings who suspend themselves in the air with two small beating wings, each about 40 centimeters long, attached to their backs. These characters are less than a

Jules Verne's *From the Earth to the Moon* was published in 1865. Breaking with literary tradition, which called for recounting such a voyage only as an imaginary undertaking, Verne based his account on an extrapolation of contemporary scientific principles. The resulting prophetic qualities of this novel are uncanny. For instance, Verne chose a launch site not far from Cape Canaveral in Florida; he also gave his readers the initial velocity required for escaping the earth's gravitation. In the sequel, *Around the Moon,* Verne correctly described the effects of weightlessness, and he even pictured the spacecraft's fiery reentry and splashdown in the Pacific Ocean—amazingly, at a site just three miles from where *Apollo 11* landed on its return from the Moon in 1969.

—ARTHUR EVANS AND RON MILLER, "JULES VERNE, MISUNDERSTOOD VISIONARY," *SCIENTIFIC AMERICAN* (APRIL 1997)

meter tall but probably have a mass of at least 20 kilograms. Could these wings suffice?

## 71. Shrunken People

Suppose someone is shrunken by some gimmick in the movies. Let's say that you suffer this consequence and are now 100 times smaller in all dimensions. Actually, there is a lot of space between the atoms and molecules of our bodies, but let's ignore any increase in repulsive forces, etc., and assume that this shrinkage can be done. What does physics tell you will be a major problem as you walk?

## 72. Spaceship Designs

The simple but effective spaceships of Buck Rogers and Flash Gordon have been superseded by flashy new designs with interesting shapes, sizes, and abilities. The advent of the space age in the 1950s brought about a heightened awareness of the practical physics characterizing a successful rocket or spaceship. Yet today, more than 50 years later, the ingenuity of the movie industry continues to defy the laws of physics. We see the latest nuclear-powered spaceships operating in space coming in for a landing on Earth (or other comparable planet) at a spaceport and then taking off for space a little while later in the same ship from the same spaceport. Why can't we do this feat with present-day space vehicles?

## 73. Warp Speed

Spaceships are known for their ability to turn on their warp drives to accelerate to speeds beyond the speed of light. Can present-day physics conceptually explain this capability?

## 74. North Pole Ice Melt

Environmental disasters have always been popular with filmmakers. In recent years, the trend has been toward disasters on a global scale because the public has become more aware of global environmental problems. If there is a global warming trend, there could be much ice melting at the poles of Earth. Some films have portrayed seacoasts being inundated by the rising water level. What would you predict for the sea level change if the ice only at the North Pole melted completely?

## 75. Lightning and Thunder

We see the flash of distant lightning and hear its thunder roll simultaneously in the movie. But we all know that the lightning flash arrives before the thunder in the real world, there being about five seconds of sound delay for each mile of distance to the lightning. Suppose you were in charge of a battle scene in a war movie. When editing the scenes of the explosions on the battlefield, how would you ensure the correct experience for the theater patron?

## 76. Explosions in Outer Space

Explosions in outer space on the big screen are magnificent to behold. Brilliant colors of stuff shooting outward in all directions, decreasing their density as the inverse square of the distance. The sound of the explosion rocks the spaceship with a thundering roar just as the light flash is first seen. Finally, bits of debris scream past. What do you think about this space physics?

H. G. Wells's 1914 novel *The World Set Free*, a speculative history of the future, contains the following sentence: "Nothing could have been more obvious to the people of the early twentieth century than the rapidity with which war was becoming impossible. And as certainly they did not see it. They did not see it until the atomic bombs burst in their fumbling hands."

How would you suspend 500,000 pounds of water in the air with no visible means of support? (Answer: build a cloud!)

—BOB MILLER, ARTIST

On November 10, 1907, the magazine section of the *New York Times* included the headline "Martians Probably Superior to Us."

# 77. Space Wars

One space battlecruiser after another shoots powerful laser beams that destroy the enemy's space battlecruiser. We see the powerful red laser beams strike the opponent, and we hear the explosion as the object blows apart. What wonderful physics can be learned here?

# 78. Security Lasers

Quite often the drama in a crime movie or an adventure movie is enhanced by having crisscrossed visible laser beams around the item to be protected from theft. The thief must avoid intersecting these beams during the escapade to steal the item; otherwise a security alarm will notify the appropriate authorities and the thief will be caught. If you were the movie's director, how would you make this scene to ensure good physics?

# 79. Bullet Fireworks

Bullets bounce everywhere. The bad guys shoot a lengthy burst of submachinegun fire as the hero runs through an industrial plant. The bullets impacting on steel railings, for example, give off bright flashes of light. This scene is a dramatic event for almost anyone watching the hero in a time of great peril. What can you say about the physics here?

# 80. Internet Gaming

For years people have been playing "live time" games over the Internet. If the game is checkers or poker, for example, each player must take his or her turn in proper order, so short delays are not a problem. Even when the game is a world-domination board game with multiple players, each player can submit moves at any time before the deadline. But many video games require simultaneous play by several players, so delays can mean life or death for a player's combatant in a shoot-'em-up type of action game. One can hear comments by some action game players that they tried to make their move but the Internet was too slow. What is the possible truth here?

## 81. Cartoon Stretching

Objects in cartoons are stretched and squeezed into amazing distortions and then released. Some of the characters suffer the same fate. When the body material of a cartoon character is being pulled, we often see the part closer to the applied force stretch first and then the rest follow with a small time delay. For example, a cartoon dog may pull on a character's leg, which we see being stretched while the torso remains normal, until finally the torso stretches, the arms stretch, and the character releases his or her handgrip from the doorway. Using some physics concepts, what can you say about the speed of sound in a cartoon character's body?

## 82. Infrared Images

In crime dramas and in adventure films the result of an infrared vision device is often reconstructed and shown in a sharp greenish or black-and-white format. We see the infrared faces of people as if they were originally color images seen normally with one's eyes, but now these color images have been converted to black and white. Is there any physics violation here in depicting the infrared images via this process?

PERIODIC CLEANUP OPERATION
Every time the solar cycle peaks, it causes Earth's atmosphere to expand and pull in low-orbiting debris, which burn up on reentry.

Jules Verne died in 1905. A memorial sculpture placed over his grave depicts Verne rising from his tomb, one arm reaching toward the stars. Some two decades later an American periodical called *Amazing Stories* —the first magazine exclusively to feature tales of science and adventure—used a representation of Verne's tomb as a logo. To describe these narratives, the publisher, Hugo Gernsback, coined the term "scientifiction," which was later changed to science fiction.
—ARTHUR B. EVANS AND RON MILLER, "JULES VERNE, MISUNDERSTOOD VISIONARY," *SCIENTIFIC AMERICAN* (APRIL 1997)

One of the big uncertainties in calculating the path of an asteroid involves how much sunlight it absorbs and then reradiates as thermal energy. Such radiation can, over the centuries, gently push the asteroid into a different orbit, much as a tiny rocket would. So, if scientists in future years should conclude that a collision looks ever more likely, then they can probably find ways to alter the asteroid's radiation pattern by dusting its surface with soot or powdered chalk or draping it with reflective Mylar. Such tinkering could be enough to nudge the asteroid safely away.

—EDITORIAL DESK, "ENCOUNTER WITH AN ASTEROID," *NEW YORK TIMES* (APRIL 8, 2002)

If you are a scientist, you're more likely to be killed in a film than a member of any other profession, including a Mafia hit man.

—CARL SAGAN

# 83. Light Sabers

Enemies dueling with light sabers have graced the silver screen for several decades now. Isn't this type of weapon the most ridiculous thing you've ever seen?

# 84. Force Fields

In battle scenes of many science fiction movies we see the baddies roll up with their giant laser guns to shoot the good guys, who are protected by a visibly transparent force field. Why do the laser beams suffer deflection at the force field?

# 85. Cold Silence of Space

In the "cold silence of space" begins many a description of space between planets. Can this statement survive a physics analysis?

# 86. Nuclear Submarine

Several movies have involved an out-of-control nuclear reactor aboard a nuclear submarine. We are told that the containment vessel is about to fail and that the best action is to move the sub several hundred meters underwater. When the explosion occurs down there, what might happen?

# 87. Plutonium vs. Uranium

Suppose you find a nuclear bomb and decide to transport the device to a safe hiding place. Would there be any difference with regard to your safety as to whether the device is made of uranium-235 or plutonium-239?

# 88. Nuclear Detonation

The threat of the detonation of a hydrogen nuclear warhead by striking one with another object such as a missile or the shrapnel from a nearby explosion lurks in a scriptwriter's creative mind for many war and adventure films. Suppose the nuclear warhead is aboard an ICBM and is struck by an interceptor missile or the warhead itself is penetrated by fast-moving BBs. What will happen?

# 89. Fabric of Space-time

Conjectures about the "fabric of space-time" and "tears in the space-time continuum" abound in science fiction movies. An entertaining 2001 film involved a protagonist who derived an equation for the time and place of a temporary tear in the fabric of space-time. Several characters jumped off the Brooklyn Bridge through the temporary space-time tear, acting as a portal to another dimension to the year 1876, and then returned through the next temporary tear by jumping off the bridge again days later. In addition, film characters have used the phrase "speed of gravity" in an ambiguous way. What can you say about the physics here?

In a sense, 1:Sun = Moon, and 1:Moon = Sun! 1:365.242 = 0.0027379, which in days is 3 minutes and 56 seconds, the difference between sidereal and solar days, while 1:27.322 = 0.0366, which in days is 52 minutes, the difference between lunar and solar days.

—ROBIN HEATH, *SUN, MOON, & EARTH*

On the average there is one catalogued satellite that falls back to Earth uncontrolled every single day and has been since the early 1960s. Most of them vaporize high in the atmosphere.

The first sentence in H. G. Wells's 1908 novel *The War in the Air* reads: "Lower Manhattan was soon a furnace of crimson flames, from which there was no escape."

# 5 | Go Ask Alice

**P**ERHAPS NO OTHER ASPECT OF TWENTIETH-century physics has captured the imagination of the general public more than the concepts of the special theory of relativity (STR). Absolute time and absolute space are forever cast aside in favor of a union of space and time into one important entity called space-time. This four-dimensional world of space-time has spawned an enormous number of conjectures about the behavior of nature. Among these conjectures are time travel, two people aging at different rates when one remains on Earth and the other travels on a space journey, the ability to see the back side of an approaching cube, and the conversion of mass into energy. As you know, the STR is based on the idea that two observers in different *inertial* reference frames must each experience physics described by the same basic laws. Even though these two inertial reference frames are moving with a constant velocity with respect to each other, the speed of light in a vacuum is the same for both observers. The important quantities in STR are the invariants. For many people, the most useful invariant is the space-time interval $\tau$, defined by $\tau^2 = c^2\, \Delta t^2 - \Delta x^2 - \Delta y^2 - \Delta z^2$. For others, the four-momentum invariant $E^2 - p^2 c^2 = m^2 c^4$ is the most useful because $E_0 = mc^2$ can be derived directly, where the mass m is a constant, the same at all speeds, all places, and all times. Many challenges in this chapter test your ability to use these invariants.

# 90. Spotlight

Can a spot of light move faster than c, the speed of light? For example, if a lighthouse light beacon spins around at very high speed, will the spot of light seen far from the beacon cut across the sky with a speed greater than $3 \times 10^8$ m/s?

# 91. Quasar Velocity

Quasars have been detected that have recessional velocities greater than the speed of light c based on the cosmological relationship for the redshift $z$, namely, $l + z = \exp[v/c]$. That is, there are quasars with $z > 3$, for example. Also, to explain the present state of the universe, the inflationary big bang model requires a faster-than-light expansion of space in the young universe. Are these examples violations of the special theory of relativity?

# 92. Spaceship Approach

A spaceship is approaching Stephanie at the relativistic speed of $v/c = 0.98974$. What does she see as the spaceship nears and then passes? Hint: for simplicity, consider a cube approaching in place of the spaceship.

# 93. Mass and Energy

A symbol of the twentieth century is the famous Einstein relation between mass and energy. Here are four possible equations: (1) $E_0 = mc^2$ (2) $E = mc^2$ (3) $E_0 = m_0 c^2$ (4) $E = m_0 c^2$. In the equations c is the velocity of light, $E$ is the total energy of a free body, $E_0$ its rest energy, $m_0$ its rest mass, and $m$ its mass.

Which of these equations expresses one of the main consequences of the STR? Which equation was first

Joseph Larmor in 1900, stimulated directly by the Michelson-Morley experiment, gave for the first time the full Lorentz transformations for coordinates and time, as well as electric and magnetic field components, and showed that the Maxwell equations remain exactly invariant under these transformations. It has long appeared a historical anomaly that Larmor's work, which preceded Lorentz's by four years, is so little known among physicists. Earlier still, in 1897, Larmor had discovered time dilation. Woldemar Voigt's paper, published in 1887, contains an early version of the Lorentz transformations that appear to be almost the same except for a scale factor.

—C. KITTEL, "LARMOR AND THE PREHISTORY OF THE LORENTZ TRANSFORMATIONS," AMERICAN JOURNAL OF PHYSICS (SEPTEMBER 1974)

Are not gross bodies and light convertible into one another?

—Isaac Newton

written by Einstein and was considered by him a consequence of STR?

## 94. Strain Gauge

A long rectangular bar of metal sits at rest in my reference frame. The strain gauge attached to its middle reads zero. Now I run in the direction parallel to the length of the bar at an enormous constant speed V at nearly light speed. I measure the bar length to determine that Lorentz-Fitzgerald contraction has occurred—that is, that the bar measures shorter than before. What should the strain gauge show?

## 95. Mass/Energy

Under certain conditions, mass can be converted into energy à la $E_0 = mc^2$. Under certain restricted conditions, energy can materialize as mass. What is wrong in these statements?

## 96. System of Particles

A system of particles is composed of $n$ freely moving particles. Is the mass of this system equal to the sum of the masses of the individual particles?

## 97. Light Propagation

Suppose Patricia is driving her car at nearly the speed of light and turns on her headlights. For simplicity in calculations, in the rest frame of an observer on the ground the light takes one second to reach the stop sign $3 \times 10^8$ meters away. This ground observer then sees the car reach the stop sign very soon after the initial light reaches the stop sign.

Patricia sees the light moving forward at $3 \times 10^8$ m/sec also, but she sees the stop sign approaching her at nearly light speed. Therefore she sees the arrival of the light flash at the stop sign and her arrival there in quick succession.

Call the initial arrival of the light at the stop sign event A and the car's arrival event B. Will the elapsed time between events A and B be the same for the driver as for the observer on the ground? No, because the ground observer sees both events occur at the same location, at the stationary stop sign, so $\Delta x = 0$. As seen by Patricia, these two events occur at two different locations separated by $\Delta x \neq 0$.

Who measures the longer time interval between events A and B? Can you provide a conceptual argument for this nonintuitive result? If the speed of the car is closer to the speed of light, how does the difference in elapsed times measured by driver and ground observer change?

# 98. Sagnac Effect

Suppose two identical clocks are in motion on Earth's Equator with constant speed v relative to Earth, one moving east and one moving west around the Equator. Do they tick at the same rate? What do their elapsed times reveal when they meet again?

# 99. Light Flashes

Suppose that a spaceship travels at constant velocity between two planets, A and B. The spaceship sends out a light flash in all directions every 10 minutes by its own clock reading. Traveling toward B, its light flashes are seen at 5-minute intervals on planet B. What is the flash interval time as seen on planet A? One of these

The notion of the dependence of mass on velocity according to $m/(1 - v^2/c^2)^{1/2}$ was introduced by Lorentz in 1899 and then developed by him and others in the years preceding Einstein's formulation of special relativity in 1905.

—LEV B. OKUN, "THE CONCEPT OF MASS," *PHYSICS TODAY* (JUNE 1989)

I'll be so happy and proud when we are together and can bring our work on relative motion to a successful conclusion!

—ALBERT EINSTEIN IN A LETTER TO MILEVA MARIĆ, HIS FUTURE WIFE, MARCH 27, 1901. JÜRGEN RENN AND ROBERT SCHULMANN, EDS., *ALBERT EINSTEIN, MILEVA MARIĆ: THE LOVE LETTERS*

A student riding in a train looks up and sees Einstein sitting next to him. Excited, he asks, "Excuse me, Professor. Does Boston stop at this train?"

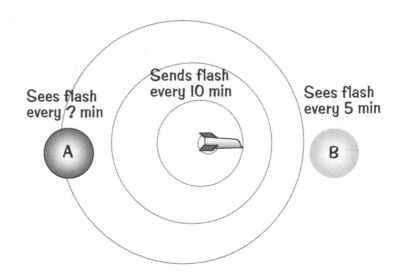

possibilities is correct: 5-minute intervals; 10-minute intervals; 15-minute intervals; 20-minute intervals.

# 100. Forces and Accelerations

In Newtonian physics, an applied contact force acting on a rigid object will accelerate the object in the same direction as the applied force. Does this behavior hold for applied contact forces in relativity physics (STR)? For example, if an applied contact force pushes on the same rigid object in the direction perpendicular to the direction of motion, will the resulting acceleration be in the direction of the applied contact force?

# 101. Uniform Acceleration

Suppose an object starts at rest with respect to the lab frame and undergoes a uniform acceleration $a'$ as measured by an observer on a spaceship moving at a uniform velocity v with respect to the lab. In Newtonian mechanics, for speeds where v << c, the velocity after $t'$ seconds in the moving frame has elapsed is

V′ = a′ t′ as measured by the observer on the moving object. This velocity is V = v + a′ t after the elapsed time *t* in the lab frame, because in Newtonian physics the clocks in the different frames run at the same rates. What is the velocity value in the lab frame when the speed is allowed to become relativistic? Can the product a*t* be greater than c in either reference frame?

# 102. Long Space Journey

Can a person go with a 1-g acceleration to a distant location 7,000 light-years away *and return* without aging more than 40 years? That is, the bathroom scale in the spaceship must show a person's correct weight for the whole journey. Is this feat within the realm of science or science fiction?

# 103. Head to Toe

Can relativitic effects make your feet age more slowly than your head?

# 104. Neutrino Mass

Since their proposed existence in the 1930s, neutrinos and antineutrinos of all three lepton families have been thought to have zero mass and travel at light speed to conserve energy and angular momentum in nuclear decays. In 1969 came the first hints that at least one type of neutrino can become another type of neutrino, and a neutrino oscillation scheme was proposed. We now know that muon neutrinos created in Earth's atmosphere can oscillate into electron neutrinos and tau netrinos before reaching an underground detector. Why cannot all three neutrino types still have zero mass?

Henri Poincaré, building on Lorentz's work but removing, at least formally, certain inconsistencies, arrived in 1905, and more fully in 1906, at the expressions $E = mc^2/(1 - v^2/c^2)^{1/2}$ and $p = mv/(1 - v^2/c^2)^{1/2}$. Einstein obtained the same relations, at the same time, on purely kinematic grounds. These are the well-tested and familiar expressions of today.

—J. DAVID JACKSON, "THE IMPACT OF SPECIAL RELATIVITY ON THEORETICAL PHYSICS," *PHYSICS TODAY* (MAY 1987)

Einstein simply postulates what we have deduced, with some difficulty and not altogether satisfactorily, from the fundamental equations of the electromagnetic field.

—H. A. LORENTZ, 1906, QUOTED IN ALBRECHT FÖLSING, *ALBERT EINSTEIN: A BIOGRAPHY*

# 105. Spaceship Collision

Two spaceships, A and B, move toward one another on courses for a head-on collision. According to an observer at rest in an inertial reference frame, both have speed $V$ along the $x$-axis. At the time of observation, spaceship A is coincident with the observer—that is, has the same $x$ value. Spaceship B is at a distance $L$ away. One would like to know how much later the collision occurs according to the observer and according to an observer aboard spaceship A.

Let us propose a solution method. According to the observer, the collision occurs when spaceship A or B travels $L/2$, half the distance between them, which requires the elapsed time $T = L/2V$. Put into a better format, three events occur:

| | | |
|---|---|---|
| Event 1: | $X_1 = 0$ | $T_1 = 0$ |
| Event 2: | $X_2 = L$ | $T_2 = 0$ |
| Event 3: | $X_3 = L/2$ | $T_3 = L/2V$ |

These same events can be specified in the inertial (primed) frame of spaceship A as:

| | | |
|---|---|---|
| Event 1': | $X_1' = 0$ | $T_1' = 0$ |
| Event 2': | $X_2' = ?$ | $T_2' = ?$ |
| Event 3': | $X_3' = ?$ | $T_3' = ?$ |

# 106. Twin Paradox

On their twenty-first birthday, Peter leaves his twin brother, Paul, behind on Earth and goes off in a straight line for 7 years on his own wristwatch time ($2.2 \times 10^8$ seconds) at 0.96 c with respect to an inertial reference frame at rest with respect to Earth, then reverses direction, and in another 7 years of his time returns at the same constant speed. Paul sees Peter's wristwatch running slower, so Peter ages $\sqrt{(1 - v^2/c^2)} = 0.28$ as much, or 1.96 years for each direction. But Peter looks back to see Paul's clock running slower than his own wristwatch, so Paul should be aging slower by 0.28 as much—that is, 1.96 years for each direction. On his return, Peter is surprised: "I *know* that I aged 14 years, but Paul should have aged only 3.92 years. Why is Paul an old man with gray hair?"

From a historical perspective, Einstein's recognition of $E = mc^2$ (where c is for "celeritas," from the Latin for "swiftness") did not quite come "out of the blue." Already in 1881, J. J. Thomson had calculated that a charged sphere behaves as if it had an *additional* mass of amount $4/3c^{-2}$ times the energy of its Coulomb field. That set off a quest for the "electromagnetic mass" of the electron—an effort to explain its inertia purely in terms of the field energy. In 1900, Poincaré made the simpler observation that since the electromagnetic momentum of radiation is $1/c^2$ times the Poynting flux of energy, radiation seems to possess a mass density $1/c^2$ times its energy density.

—WOLFGANG RINDLER,
*RELATIVITY: SPECIAL, GENERAL,
AND COSMOLOGICAL*

# 6 | Start Me Up

Engineering physics is really applied physics, but more general, with social, political, financial, and aesthetic issues to be considered that are often beyond the immediate concerns of the applied scientist. We have included the ability to understand the microscopic behavior of atoms and their components in solid and liquids, a knowledge that has begun to reap huge benefits in improving the materials and devices around us. In fact, we have entered the era of ingenious devices and designer materials. A very small sampling of the vast array of these advances is included in the challenges and puzzles considered in this chapter.

## 107. Air-Driven Automobile Engine

Can a normal four-cylinder gasoline engine actually operate on compressed air instead of gasoline as its energy source?

## 108. Coin Tosses

The behavior of many systems and materials can be better understood by considering the random walk of particles in the system. To get some "feeling" for a random walk, consider the following exercise. Divide a group of people into two groups. Have each individual in one group toss a fair coin 256 times and write down in sequence the outcome of each toss. Have each individual in the other group write down what they would *imagine* a typical sequence of 256 random tosses to be but not actually do the tossing. Collect all the papers and mix them up thoroughly. Can you determine with reasonable accuracy which sets of data were obtained experimentally? How accurate should your selection be?

## 109. More Coin Tosses

Suppose we are really ambitious about tossing a fair coin. Indeed, suppose we toss a fair coin 1000 times, and for each head we step one unit distance radially away from a lamppost, and for each tail we step back radially the same unit distance. About how many times would you expect to be at the lamppost?

## 110. Brownian Motor

In his famous lectures, physicist Richard Feynman discussed the impossibility of violating the second law of thermodynamics by a *ratchet* mechanism. The simplest model for a ratchet is an overdamped Brownian particle in an asymmetric but spatially periodic potential (with

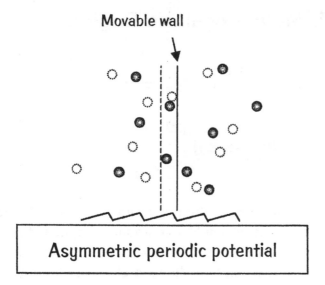

Movable wall

Asymmetric periodic potential

asymmetry and period *L*). Due to the fluctuating force caused by the pushing molecules of the surrounding fluid or gas, the Brownian particle may overcome the potential barrier moving to the left or to the right. The probabilities for both directions are equal, and on average the particle does not move. Hence building a motor that turns thermal energy into mechanical work from a *single* heat bath is impossible.

But the ratchet can be turned into a so-called Brownian motor that seems to violate the second law of thermodynamics. The idea is to turn the ratchet potential on and off periodically. Under certain circumstances, this action may yield directed motion even against an applied force *f*. Indeed, this device does work. (No pun intended!)

Recall that a perpetuum mobile of the first kind violates the law of conservation of energy, while a perpetuum mobile of the second kind uses the "free" energy around us in the form of heat—that is, random thermal motion of molecules and atoms—to run an engine without fuel. Why isn't a Brownian motor a perpetuum mobile of the second kind?

## III. Magnetocaloric Engine

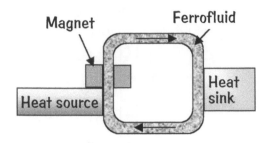

A ferrofluid is a fluid containing small magnetic particles that respond to an applied magnetic field, so a ferrofluid becomes magnetized in the presence of the magnet. The diagram shows a closed tube loop containing a ferrofluid, a heat source, a strong magnet, and a heat sink all working together to act as an engine transporting the ferrofluid around the closed loop. Its thermal efficiency approaches the efficiency of a Carnot cycle, so demands for this device should increase. Exactly how does this engine maintain the fluid movement around the loop? Can a solar heating system operate in this way?

# 112. Magnetorheological Fluid

In a beaker is 250 milliliters of corn oil to which has been added about 0.5 kilogram of iron filings about 1 millimeter long. The mixture is stirred thoroughly and a strong horseshoe magnet is brought up to straddle the beaker. The iron filings align with the magnetic field as expected to magnetize the fluid mixture. What other physical property of the fluid changes?

# 113. Binary Fluids

The two possible phase diagrams show the miscible and immiscible phases of a binary fluid, a mixture of two

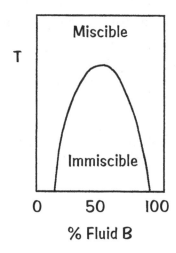 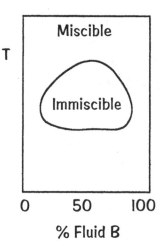

kinds of fluid, in a plot with axes of temperature versus concentration. For example, coffee and cream are miscible at room temperature but oil and water are not.

Consider the 50 percent mixture in each phase diagram and start at a high temperature in the miscible phase. The diagram to the left reveals that the binary fluids become immiscible upon being cooled, while the diagram to the right tells us that the fluids become immiscible as the cooling proceeds but that even further cooling brings back the miscible phase. Can both phase diagrams represent a real binary fluid, or is one false?

## 114. Baseball Bats

Hitting a baseball well is not easy. Even professional baseball players have difficulty consistently making solid contact with a pitched baseball. Once hit, the distance of flight of the ball is determined by its initial velocity—that is, the initial speed and direction—which depends on how hard the ball has been hit by the bat. All other factors being held constant, the initial velocity can be said to depend on the speed of the bat just before collision. A quicker swing would mean a faster bat

A technician named Richard Woodbridge III coined the phrase "acoustic archaeology" in the August 1969 issue of *Proceedings of the I.E.E.E.* Woodbridge theorized that there were many occasions when sound might innocently get scooped out of the air and preserved. For example, when an ancient potter typically held a flat stick against a rotating pot, he was accidentally (and crudely) recording into the clay the sounds around him. Woodbridge wrote about experiments he performed pulling basic noises off a pot. Another experiment involved setting up a canvas and then talking while making different brush strokes, hoping to record a spoken word in an oil portrait. In this fashion, for instance, the word "blue" was pulled off a blue paint stroke.

—JACK HITT, "EAVESDROPPING ON HISTORY," *NEW YORK TIMES MAGAZINE* (DECEMBER 3, 2000)

speed during the collision to add distance to each hit and also allow the batter more time to judge the pitch.

There have been proposals to put shallow, pea-sized depressions—dimples—in the surface of a baseball bat to allow a greater swing speed. Another sports object, the golf ball, already is made with dimples on its surface. How would these dimples affect the bat's swing speed?

## 115. Old Glass

In old castles and houses in Europe can be found windows with old glass in which many of the panes are slightly thicker on the bottom than at the top. What are some possible reasons for this result, and what is the most likely reason?

## 116. Ferromagnetism

Why are so few substances ferromagnetic, yet practically all materials exhibit paramagnetic behavior?

## 117. Coupled Flywheels

Conservation of angular momentum does not always help in understanding the behavior of rotating devices. The diagram shows two flywheels, 1 and 2, of moments of inertia $I_1$ and $I_2$ mounted on parallel horizontal shafts along with pulleys of diameters $D_1$ and $D_2$. The belt is slack at first, and the two flywheels are

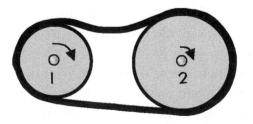

running at angular velocities $\omega_{10}$ and $\omega_{20}$. Suddenly the belt is tightened. One can write out the torque equations and the angular momentum equation to get the relation $I_1 \omega_1 + I_2 \omega_2 = k - (N - 1) I_1 \omega_1$. Here, k is a constant of integration and $N = D_2 / D_1$, the ratio of pulley diameters. When $N = 1$, angular momentum is conserved. If $N \neq 1$ and $\omega_1$ changes, the angular momentum is not conserved! Why not?

## 118. Superconductor Suspension

A popular physics demonstration since the late 1980s involves floating a small piece of high-temperature superconductor, such as yttrium barium cuprate ($YBa_3Cu_3O_7$), over a strong permanent magnet. The levitation is easy to see, and the suspended superconductor rectangular solid spins rapidly about its long axis. The demonstration is done by first cooling the superconductor in liquid nitrogen and then using tongs to place the piece in the air above the permanent magnet. The repulsive force between the magnet and the superconductor is a demonstration of the Meissner effect. Or is it?

## 119. Nanophase Copper

The hardness and strength of a metal are measured by studying its deformation in response to an applied force. A metal is deformed when its crystalline atomic planes slide over each other. An analogy may be the bump in a rug that can be pushed across the floor. In other words, a dislocation in a plane of atoms is moved until a barrier is reached, such as a grain boundary, where the micron-sized grains are differently oriented.

One interesting advance in metal technology is the ability to assemble nanometer-size clusters of atoms in grain sizes of less than 100 nanometers in diameter instead of having the micron-size grains found in a typical metal. A graph of hardness versus grain size is shown.

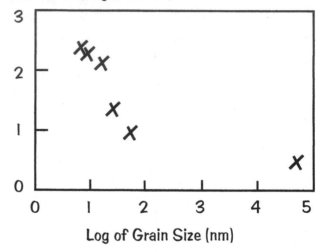

With grain sizes averaging about 10 nanometers, this nanophase copper metal has a hardness more than three times the hardness of normal copper metal. Why?

## 120. Head of a Pin

What is the smallest amount of charge that can sit on the head of a pin? Some people say that the smallest nonvanishing amount of charge should be +e or –e, where e is the fundamental unit of electrical charge. What do you say?

## 121. Coulomb Blockade

The tunnel junction is a conductor-insulator-conductor device. Suppose a very small tunnel junction is operated at very low temperatures so that thermal fluctuations do not contribute to electron tunneling across the junction. Now connect the tunnel junction to a source of

Jean Buridan (ca. 1295–ca. 1358), rector of the University of Paris in 1327, in his impetus theory introduced the prescient notion that the true measure of the motion of an object was not speed alone, but the product of speed and quantity of matter (*quantitas materiae*). In an anticipation of Newton's first law of motion, he maintained that once the initial impetus was supplied, motion continued indefinitely. The spheres of heaven, for instance, having been put in motion by God, continued so and required no constantly working angels to keep them moving.

—Isaac Asimov, *Asimov's Biographical Encyclopedia of Science and Technology*, 2nd rev. ed.

constant electrical charge. Will the flow of current across the junction be steady?

# 122. Deterministic Competition

Consider a simplified system, one that can be described by $N_t$ objects at time $t$. For example, one could consider the number of grasshoppers on the plains of Africa, or on some small plot of land. Let there be competition between the growth processes and the decay processes so that the number of objects at time $t + 1$ is $N_{t+1} = N_t \exp[r (1 - N_t)]$, an exponential growth relationship. This equation is deterministic, for $N_t$ determines $N_{t+1}$ unambiguously. One can think of $r$ as a measure of the ratio between growth and decay. Numerous mechanical, hydrodynamic, chemical, and electrical systems can be approximately modeled by this relationship.

How does the number of objects behave with elapsed time? If $N_t = 1$, then $N$ remains 1 forever. In the general case, we can determine $N_t$ as $t \to \infty$ to find out whether $N$ approaches the equilibrium value 1. For instance, let $r = 1$ and begin with $N_0 = 0.5$, and calculate with a calculator or personal computer. Now try different values for $r$. What behavior do you predict?

# 123. Two Identical Chaotic Systems

A chaotic system exhibits a sensitivity to initial conditions and will evolve rapidly and deterministically toward different end states if begun in slightly different states. Although the chaos is unpredictable, each possible outcome is deterministic—that is, an orderly behavior.

Consider two identical chaotic systems isolated from each other. They will quickly fall out of step because any slight difference between them would be magnified. Assume that these systems have several

Descartes regarded the conservation of momentum (quantity of motion) as divinely ordained. He wrote: "[God] set in motion in many different ways the parts of matter when He created them, and since He maintained them with the same behavior and with the same laws as He laid upon them in their creation. He conserves continually in this matter an equal quantity of motion."

Earth is gradually slowing down; the day is about 16 milliseconds longer now than it was 1,000 years ago. This slowing is due largely to frictional tidal effects of the Moon on Earth's oceans.

parts and that at least one of the parts is stable—that is, subjected to a perturbation, the part's behavior changes a little but settles back to its normal operation. Now drive both systems with the same chaotic signal applied to the same stable part. Can the two systems be synchronized?

## 124. Tilley's Circuit

This electrical circuit near the permanent magnet has two ideal switches and a galvanometer. When switch A is closed and switch B, on the right, is opened, there is a large change in the magnetic flux in the galvanometer circuit. What do you predict the galvanometer response will be?

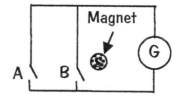

## 125. Thermal Energy Flow

If two identical bodies at different temperatures are in contact, thermal energy will always flow from one to the other in such a direction as to increase the total entropy. In which direction will this flow be? That depends on two factors, the amount of energy and entropy the two bodies already contain. The second law of thermodynamics implies that thermal energy must flow toward the region of lower temperature—that is, each unit of thermal energy acquires *greater* disorder as it moves into the cooler region. Why?

# 126. Cadmium Selenide

When atoms are arranged in nanometer-size clusters of diameters from less than 100 nanometers to as large as 700 nanometers, interesting optical properties can be demonstrated. For example, nanophase versions of pure cadmium selenide can be made almost any color in the spectrum simply by changing its cluster size. Indeed, some types of lipstick are made in many different colors even though the predominant light-scattering molecule is the same in all color versions. What is the physics here?

# 127. Optical Solitons

A light pulse is a continuum of optical carriers of different frequencies. Optical media are dispersive, so these carriers in the light pulse travel at different velocities, causing the energy to spread over time and distance. In addition, there is the optical Kerr effect, which "instantaneously" increases the refractive index of the medium by an amount proportional to the optical power. Can one use these two effects—dispersion and the Kerr effect—to ensure that a light pulse retains its integrity while traveling thousands of kilometers through an optical fiber?

# 128. Ceramic Light Response

Certain ceramic materials will change their shape upon exposure to light. What is the physics here?

# 129. Random Movements

Supposedly, research has revealed that random movements help explain how a tightrope walker stays aloft, for instance. If understood, robotics engineers could

Is Galileo a beneficiary of the Matthew effect? The latter is a term introduced in 1968 by Robert K. Merton (1910–2003), a U.S. sociologist of science, that refers to the disproportionately great credit given to eminent scientists for their contributions to science, while relatively unknown ones tend to get disproportionately little for their occasionally comparable contributions. The term derives, of course, from the Gospel according to Matthew (13:12 and 25:29). In the New King James Version the passage reads: "For whoever has, to him more will be given, and he will have abundance; but whoever does not have, even what he has will be taken away from him." Recognition tends to go to those who are already famous. In Galileo's case, both Philoponus in the sixth century and the Belgian–Dutch scientist Simon Stevin in 1586 performed the key experiment of dropping two different weights simultaneously and observed that they struck the ground at the same time—the experiment that today seems indissolubly, if incorrectly, wedded to the name of Galileo.

make their machines more stable by injecting a little noise into their systems. And persons having difficulty walking may be able to let some noisy vibrating shoe soles help them walk confidently again. What could be the physics here?

# 130. Gravitational Twins

Engineering physics involves the transport of people and materials in space as well as practical applications here on Earth. So consider a pair of twins in free fall. Imagine that one twin is in circular orbit around a star and that her sister is shot out from this circular orbit location on a radial orbit—that is, the traveling twin will fall back to meet up with the stay-at-home sister in circular orbit. For simplicity, let them meet after an integral number of revolutions around the circular orbit for the one left behind.

Any clock system in a gravitational potential, such as the clocks in the GPS system here on Earth, depends on two relativistic effects on the clock rate: (1) a clock ticks slower closer to a massive object than when far away, and (2) the faster-moving clock ticks slower than the slower-moving clock.

Initially, the clock rates of the twins are the same because they start out in the same circular orbit at the same radial distance from the star. The traveling twin moves away from the star along the radial line, all the while slowing down and eventually coming to a momentary stop and returning with ever-increasing speed until rejoining her sister in orbit. So on average, the traveling twin experiences a smaller amount of gravitational time dilation and a smaller amount of speed time dilation than her stay-at-home sister. Therefore the traveling twin returns home *older* than her sister, because her clock ticked faster on average. What do you think?

# 131. Photon Engine

The ideal Carnot heat engine converts heat to work without the engine itself being a source of any work. The reversible closed Carnot cycle consists of two isothermal (constant temperature) processes and two adiabatic (no external exchange of thermal energy) processes. No heat engine operating between two temperatures can be more efficient than a Carnot cycle.

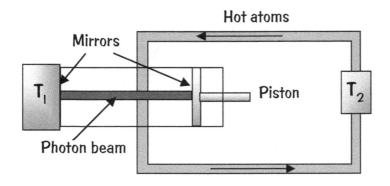

But Carnot could be wrong. The challenger is the new "quantum Carnot engine," in which the radiation pressure from photons drives a piston in an optical cavity. The inward-facing surface of the piston is mirrored and the other cavity mirror is fixed in place while exchanging thermal energy with a heat sink at temperature $T_1$. A second heat bath at a higher temperature, $T_2$, provides the source of thermal energy for the photons.

This source of thermal energy is a stream of hot atoms, which flows into the optical cavity and exchanges thermal energy with the photons through emission and absorption processes. These atoms exit the cavity at a cooler temperature and are reheated to

Nikola Tesla (1856–1943), the Serbian-born inventor of the first practical alternating-current dynamo and power transmission system, was known for unusual powers of visualization. He was able to construct, modify, and even operate his imaginary devices, purely by visualizing them. He wrote in "My Inventions" (*Electrical Experimenter*, 1919), "It is absolutely immaterial to me whether I run my turbine in thought or test it in my shop. There is no difference whatever, the results are the same. In this way I am able to rapidly develop and perfect a conception without touching anything. When I have gone so far as to embody in the invention every possible improvement I can think of and see no fault anywhere, I put into concrete form this final product of my brain. Invariably my device works as I conceived that it should, and the experiment comes out exactly as I planned it. In twenty years there has not been a single exception."

The Earth is a somewhat irregular clock. Some years the length of the day is found to vary by as much as one part in 10 million, or three seconds in a year of 31.5 million seconds. In addition, there are also seasonal fluctuations of a few milliseconds per year. In the winter the Earth slows down, and in the summer it speeds up. Think of the Earth as a spinning skater. During the winter in the northern hemisphere, water evaporates from the ocean and accumulates as ice and snow on the high mountains. This movement of water from the oceans to the mountaintops is similar to the skater's extending her arms. So the Earth slows down in winter; by the summer the snow melts and runs back to the seas, and the Earth speeds up again. This effect is not compensated by the opposite effect in the southern hemisphere because most of the land mass is north of the equator.

—James Jespersen and Jane Fitz-Randolph, *From Sundials to Atomic Clocks*

$T_2$ in a second cavity, to be reinjected into the first cavity for the next cycle of the quantum Carnot engine.

Therefore, the quantum and classical Carnot engines operate in the same way as a closed cycle of two isothermal and two adiabatic processes. However, in its simplest form, when each bath atom is treated as a two-state system, the quantum Carnot engine cannot extract work from a single heat bath. Why not? Will the engine work if each bath atom is a three-state system?

# 7 | A Whole New World

ATOMIC PHYSICS BEGAN IN THE 1840s WITH the identification of the emission lines of hydrogen and of other atoms and ions in laboratory sources and in the solar spectrum. In the early 1900s, the Bohr-Sommerfeld model of the atom was the paradigm, but numerous problems with its predictions existed that were finally resolved with the advent of quantum mechanics in 1925. The electron in the atom occupies particular quantized energy states of unequal energy spacing, and selection rules based on conservation of energy and angular momentum dictate which jumps between states to an available final state are allowed. In addition to a spontaneous electron jump to a lower energy level with the emission of a photon, external photons with the correct energy can stimulate the atomic absorption or emission of photons. The eventual application of quantum mechanics to the simple molecules proved very successful, if not challenging, and today faster computers continue to calculate the properties of atoms, inorganic and organic molecules, and very large biomolecules such as DNA and proteins. Enormous progress has also been made in understanding the fundamental properties of condensed matter of fluids and solids such as crystals, ionically doped materials, plastics, pseudocrystals, and so on. Our lives are becoming more dependent on the practical devices arising from this great endeavor called molecular design. The challenges introduced in this chapter are but a small sample of the wide range of possible problems.

## 132. Grain of Sand

If the atoms in a grain of sand were laid out side by side in a line, approximately how long would the line be?

## 133. Forensics

Historically, paintings could be verified with reasonable assurance of authorship by experts who knew the brushstrokes and color and paint choices of the artist as well as the overall style and character of the subjects. However, in some cases, fraudulent artworks have been successfully passed as genuine. New techniques for assessing all types of artwork are always needed, and the scientific community has been answering the call. One scientific technique for checking the authenticity of old paintings uses laser lights. How might this feat be accomplished?

## 134. Doppler Elimination?

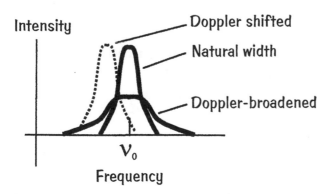

When an atom emits or absorbs a photon, there is always a recoil of the atom and a Doppler shift in the photon frequency. Is it possible to have recoilless atomic emission or absorption?

Both Niels Bohr and his wife had a similar response to religion: Margrethe has written about Niels': "There was a period of about a year . . . [he was] 14 or 15 . . . where he took it all very seriously; he got taken by it. Then suddenly it was all over. It was nothing for him." About her own feelings, Margrethe reported: "You know it was often at that age . . . that one got very religious and would listen to the minister about confirmation. Then it all dissolved. And for me it was exactly the same; it disappeared completely."

—LÉON ROSENFELD, "BOHR, NIELS HENRIK DAVID," DICTIONARY OF SCIENTIFIC BIOGRAPHY, VOL. 2

# 135. Light Tweezer

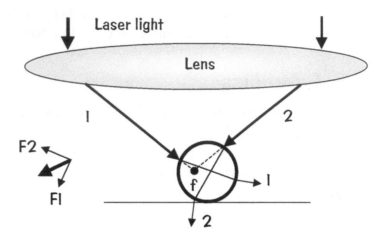

In science fiction movies we often see light beams shot out from handheld light guns supplying a tremendous impulse to knock over an enemy storm trooper approaching along the direction of the beam. By Newton's third law, the light gun itself should have experienced an equivalent recoil! We know that a ray of light has energy and linear momentum, so its impingence on any surface will produce a slight backward movement of that surface. However, we would like to know whether a light beam can be used to physically move a tiny object, such as a small one-celled animal, in a direction *perpendicular to* the beam.

# 136. Fluorescent Lights

The gas plasma inside a fluorescent tube emits mostly ultraviolet radiation and very little visible radiation. Electrons are captured by the ions and jump down to lower energies, emitting a characteristic UV photon for each fluorescence jump. Why are fluorescent tubes so much more efficient in producing *visible* light than incandescent lamps?

Why can the light from some fluorescent lights be dangerous to your health? Could it be that they emit some UV? Are there types of fluorescent tubes that are better for human working environments? Are they better because they do not emit in the UV?

## 137. Phase Conjugation Mirror

Can a light wave pass through a disturbing medium, be distorted, reflect off a special mirror, and return to the source *undisturbed*?

## 138. Stationary States

In the Bohr model of the hydrogen atom, the angular momentum for the orbital motion of the electron of mass $m$ at distance $r$ is quantized in integral units of Planck's constant h—that is, assuming the proton position to be fixed, $mvr = nh/2\pi$, where $n$ is an integer and $v$ the electron velocity. Using $mv = h/\lambda$, de Broglie was able to derive Bohr's quantization rule and $n\lambda = 2\pi r$. If $f_1$ and $f_2$ are the frequencies of the Bohr orbital motion of the electron in energy states $E_1$ and $E_2$, then if an electron jumps down from state 2 to state 1, why isn't the energy of the emitted photon the difference energy $hf_1 - hf_2$?

## 139. Angular Momentum

In classical calculations, the quantity that often appears in the result is the square of the angular momentum $J^2$. One can often guess at the correct quantum mechanical formula by replacing $J^2$ by $j(j + 1)\,h^2/4\pi^2$, where $j$ is the z-component of the angular momentum and h is Planck's constant. Why is the square of the angular momentum in quantum mechanics proportional to $j(j + 1)$ instead of just $j^2$?

which still remain untranslated from the Latin. To get rid of the "spooky action at a distance," he introduced the proposition that atoms have no size; they are geometrical "points of force" that in turn create fields of force, an idea later elaborated on by Faraday. Moreover, he suggested that all these atomic forces along with gravity, must be aspects of one all-encompassing universal force, an eighteenth century version of the "theory of everything!"

—ADAPTED FROM LESLIE HOLLI-
DAY, "EARLY VIEWS ON FORCES
BETWEEN ATOMS," *SCIENTIFIC
AMERICAN* (MAY 1970)

## 140. Kinetic Laser

A traditional laser involves the stimulated downward electron transition in an atom in a background "sea of photons," with the emission of a characteristic photon matching in fequency and momentum the stimulating photons. This stimulated emission process was predicted by Einstein. In 1951, J. Weber at the University of Maryland was the first to calculate the operating principles of the ammonium maser and laser. However, as the story goes, upon asking for research monies to build the maser, a few hundred thousand dollars from the university, he lost out to the athletic department's request for money to build up the Maryland football program.

The first operating ammonium maser was subsequently built by C. Townes in 1954, and the first operating device lasing in the optical part of the spectrum was built in 1960 by T. H. Maiman. Laser action first in the microwave region is no coincidence, for spontaneous emission is proportional to the cube of the transition frequency, and being extremely small in this part of the spectrum, can be neglected compared to stimulated emission and absorption.

Among the more exotic lasers is the kinetic laser, which is an "exploding" material that emits light and X-rays. In its simplest form, the material would be a foil of a single element such as copper that is exploded by focusing powerful laser pulses on it. How does this type of laser produce coherent laser light?

## 141. Noninversion Laser

For decades, lasers have been explained as the result of an inverted population of states with stimulated emission of photons in a high-Q cavity. However, lasers can be made without an inverted population. Can you explain how this type of stimulated emission process works?

## 142. X-ray Paradox

The index of refraction $n$ gives the ratio $c/v$, the speed of light in vacuum to the speed of the electromagnetic wave in the material. Window glass, for example, can have an index of about $n = 1.5$ for visible light, with a slight variation in the index with the color of the light. A paradox arises with X-rays because they have an index of refraction value *less than one* in crystals! What does this behavior mean?

## 143. Benzene Ring

The benzene molecule is a ring of six carbon atoms, each C atom having one H atom attached. There is a mystery about the energy contained in this molecule. The benzene ring can be broken up into pieces, and chemists have measured the energies associated with the pieces and with the single bonds and the double bonds by studying ethylene and so on. The expected total energy can be calculated from these data, but the actual total energy of the benzene ring is much lower, telling us that the carbon atoms are much more tightly bound. Therefore, the bond picture would make the benzene ring easily susceptible to chemical attack, yet the molecule is quite resilient to breaking up.

Using the Schrödinger equation by considering each carbon atom on this ring as the potential home for a single electron, one can calculate the possible energy levels for the benzene ring. Why does this method of calculation work?

Why do all FM radio stations end in an odd number? FM radio stations all transmit in a band between 88 MHz and 108 MHz. Inside that band, each station occupies a 200 kHz slice, and all of the slices start on odd number frequencies. This is completely arbitrary. In Europe, the FM stations are spaced 100 kHz apart, and their frequencies can end on even or odd numbers.

When Einstein registered for the draft in Switzerland at the age of 22, his height was recorded as five feet seven and a half inches. His contemporaries regarded him as tall. By way of comparison, Isaac Newton is thought to have been about five feet five inches tall.

—ADAPTED FROM BARRY PARKER, *EINSTEIN: THE PASSIONS OF A SCIENTIST*

# 144. Graphite

Atoms in a crystal make a regular array if there are no dislocations. Most pure single-element crystals have a cubic or a diamond crystal structure, with all orthogonal directions showing the same structural spacing. Even for a pure element substance, however, the spacing may be different in different directions. For example, take carbon atoms, which probably are components of more than 75 percent of all known compounds. In diamond they have the same structure in all orthogonal directions, but in graphite the third direction is definitely quite different than the other two directions, which define a plane of hexagonal carbon rings. How can this third direction be so different in an originally nonbiased environment?

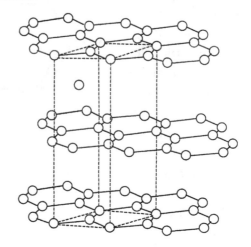

# 145. Ozone Layer

We've heard so much in the past few decades about the ozone layer in the upper atmosphere and its possible demise. Yet ozone is only a minor greenhouse gas, far behind carbon dioxide, HOH vapor, and methane in overall importance. So why is there all this fuss over the ozone layer?

## 146. Greenhouse Gases

Why are the greenhouse gases carbon dioxide, HOH vapor, and methane important for human survival on Earth? If they are good for our existence, shouldn't having more carbon dioxide, etc., in the atmosphere be encouraged?

## 147. LED vs. LCD

An LED is a semiconductor device that emits visible light when an electric current passes through it. The light is not particularly bright and usually monochromatic, occurring at a single wavelength. The LED light output range is from infrared and red to blue-violet. The LCD is a type of display used in digital watches and many portable computers that utilizes two sheets of polarizing material with a liquid crystal solution between them. An electric current passed through the liquid causes the crystals to align so that light cannot pass through, each crystal acting like a shutter, either allowing light to pass through or blocking the light.

What is the difference in energy requirements in the operation of a light-emitting diode (LED) and a liquid crystal display (LCD)? After all, they both require energy to operate. And how is a plasma display different from both of them in its energy requirements?

## 148. Sonoluminescence

Sound energy is converted directly into light energy by a phenomenon called sonoluminescence. Discovered in the 1800s, the process lay dormant for more than 100 years, only to experience a revival in the 1990s. How does one convert a small amount of sound energy into a brief but brilliant flash of light?

The special theory of relativity predicts that, for an observer moving at the speed of light, distance traveled shrinks to zero while time slows to a standstill. Thus, as far as the light itself is concerned, it does not travel any distance, and takes no time to do so. As Gilbert Lewis showed back in 1926 (*Nature*, vol. 117), from light's point of view the Universe is so "bent" that there is no separation between the point of emission of light and its point of absorption. . .

If light does not experience itself to have traveled any distance, it does not need a vehicle or mechanism by which to travel. . . . It is only in our frame of reference—the frame of observers with mass who move at sub-light speeds—that light appears to travel through space and time; and only in that frame does the question of whether it is a wave, a particle or both arise.

—Peter Russell, "Here Is There," "Letters," *New Scientist* (November 23, 1991)

# 149. Siphoning Liquid Helium

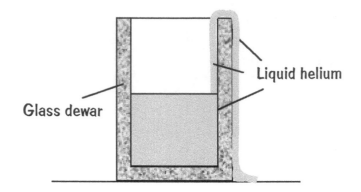

Liquid helium

Glass dewar

Liquid helium can crawl up the wall of its container without any additional help. How is this feat accomplished?

# 150. Quantized Hall Effect

The Hall effect was discovered by Edwin Hall in 1879. ". . . A charged particle moving in a magnetic field feels a 'Lorentz' force perpendicular to its direction of motion and the magnetic field. As a direct consequence of this Lorentz force, charged particles will accumulate to one side of a wire if you send current through it and hold it still in a [perpendicular] magnetic field. . . ." When the transverse voltage is measured at a fixed current, the Hall resistance is measured and increases linearly with an applied magnetic field.

The conduction electrons in a solid behave like a gas of electrons. So the discovery of the *quantized* Hall effect in 1980 by von Klitzing and his research group when he was investigating the conductance properties of two-dimensional electron gases at very low temperatures and high magnetic fields was a surprise. What is the physics behind this quantized Hall effect?

## 151. Integrated Circuits

As integrated circuits (ICs) become crowded with more semiconductor devices and internal connections, one wonders how they will be connected to the external world. We know also that cosmic rays and other particle radiation from the environment already disrupt some of the operations by random destruction, and these effects will become worse as the scale diminishes. However, neither connection to the external world via gold wires of any size nor the background particle radiation is the major problem today. What is?

## 152. Atomic Computers?

Atoms are busy collections of electrons and nuclear particles that are ever changing their positions in a random dance. In contrast, information storage requires stable states over reasonable time intervals. Can information be stored on individual atoms in their restless world?

## 153. X-ray Laser?

We know that there exist free electron X-ray lasers that have an electron moving past a rippled surface and emitting X-rays, as well as X-ray laser sources based on

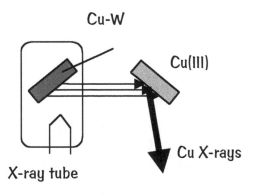

Cu-W

Cu(III)

Cu X-rays

X-ray tube

Planetary atomic models were already popular a few years before Rutherford's proposal. The most elaborate attempt was that of Hantaro Nagaoka, whose "Saturnian" model was published in 1904. Nagaoka's model was astronomically inspired, in the sense that it closely relied on Maxwell's 1856 analysis of the stability of Saturn's rings. The Japanese physicist assumed that the electrons were placed uniformly on rings moving around the attractive center of a positive nucleus. Nagaoka's calculations led to suggestive spectral formulas and a qualitative explanation of radioactivity. The model was, however, severely criticized, and disappeared from the scene only to reappear in an entirely different dressing with Rutherford's nuclear theory.

—HELGE KRAGH, *QUANTUM GENERATIONS: A HISTORY OF PHYSICS IN THE TWENTIETH CENTURY*

There was a time when physics and philosophy were allied disciplines. However, Niels Bohr's three long historic papers on the structure of the hydrogen atom were published in 1913 in *Philosophical Magazine*. The journal, first published in 1798, was at that time accepting articles from most branches of science. This alliance began to break up at the end of the nineteenth century. Today, even though the word "philosophical" persists in its title, it is devoted primarily to condensed-matter physics.

Things that cannot go on forever don't.

—Anonymous

plasmas such as the kinetic laser. However, an X-ray laser with a wavelength of about 1 Å or 0.1 nanometer or less that can be operated on a tabletop would be convenient and would be able to resolve details down to nearly 1 wavelength. The uses in physics and medicine are expected to be many.

A very interesting tabletop device is the working monochromatic X-ray source shown in the illustration that emits very intense, narrow beams at the Cu 1.54 Å characteristic emission line known as $K\alpha_1$. There is a special bimetal X-ray tube source of Cu-W that emits X-rays from both metals upon bombardment with high-energy electrons in the standard way. These X-rays exit the tube and then Bragg scatter in an *external* Cu crystal to produce a very narrow, intense beam of Cu $K\alpha_1$ X-rays. The first surprise is the enormous line intensity at a single wavelength, and the second surprise is that no Cu $K\alpha_2$ X-rays appear in the output from the external crystal. How does the external crystal affect the X-ray beam? Is this device an X-ray laser or a super-radiant X-ray source?

# 154. Bose-Einstein Condensate

A Bose-Einstein condensate is a new form of matter made at the coldest temperatures in the universe. Essentially the condensate is a collection of identical atoms behaving as one entity. How do the individual atoms lose their self-identity?

# 155. Quantum Dots

Quantum dots are crystals containing only a few hundred atoms and when illuminated with UV light, for example, will fluoresce at only one specific wavelength of light. Why does the dot emit only one wavelength of light when excited?

# 8 | Chances Are

QUANTUM MECHANICS (QM) ORIGINATED IN 1925 as a theory to understand the internal behavior of the hydrogen atom. Since then, QM has evolved to encompass the behavior of practically everything. In its most rudimentary version, QM is based on three fundamental rules. The main idea of QM is not quantized energy and quantized angular momentum, for the classical physics of strings, tubes, drumheads, and so on, involve quantized states of energy and angular momentum.

The heart of QM is the coherent superposition of states, as given in rule 2 below. From *The Feynman Lectures on Physics,* the three fundamental rules of QM are:

1. The probability $P$ of an event in an ideal experiment is given by the square of the absolute value of a complex number $\psi$, which is called the probability amplitude (or wave function):
$$P = |\psi|^2.$$

2. When an event can occur in several alternative ways, the probability amplitude $\psi$ for the event is the sum of the probability amplitudes $\psi_1$, $\psi_2$, $\psi_3$ . . . , for each way considered separately; that is, there is superposition and interference:

$$\psi = \psi_1 + \psi_2 + \psi_3 + \ldots$$
$$P = |\psi_1 + \psi_2 + \psi_3 + \ldots|^2$$

3. If an experiment is performed (or could be done) that can determine whether one or another alternative is actually taken, the probability of the event is the (classical) sum of the probabilities for each alternative; that is, the interference is lost:

$$P = P_1 + P_2 + P_3 + \ldots$$

We have no knowledge about a more basic mechanism from which these rules can be deduced. Numerous tests have verified their fundamental validity over and over. You will need to apply them in the challenges.

# 156. Schizophrenic Playing Card

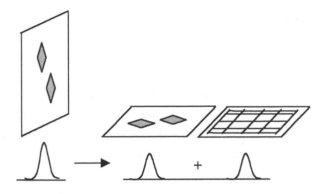

An ideal playing card stands perfectly balanced on its edge. According to the rules of quantum mechanics, this card will fall in both directions at once! That is, the final state of the card is the superposition of the two alternative falling directions, with $\psi_1$ for left and $\psi_2$ for right. The card's wave function changes smoothly and continuously from the balanced state to the mysterious final state $\Psi = \psi_1 + \psi_2$ with two alternatives that seem to have the card in two places at once. Why haven't we seen this happen in the everyday world around us?

# 157. Schrödinger's Cat

In one version of the famous Schrödinger cat *gedanken* experiment, a healthy cat is placed inside an ideal cat playroom that is isolated from the rest of the world whenever the door is closed. Inside is one deadly object left by mistake. The door is closed. After some time elapses, one wonders whether the cat is alive or dead, the two classical possibilities. Rule 2 of QM tells us, however, that the state of the cat is $\Psi = \psi_1 + \psi_2$, where $\psi_1$ means alive and $\psi_2$ means dead. So QM requires us to consider the cat as being alive *and* dead simultaneously! However, you are curious. You push a button

For centuries, Britain and its colonies rang in the New Year on March 25, Annunciation Day, when according to the biblical account the angel Gabriel announced to the Virgin Mary that she would bear the child of God. March 25 is nine months before Christmas.

—Duncan Steel, *Marking Time: The Epic Quest to Invent the Perfect Calendar*

If you stood on the moon's near side, you would see the Earth suspended against the stars more or less in the same direction with respect to your horizon—never rising or setting. But the Earth as seen from the moon would exhibit phases over the course of a month, just as the moon does as seen from Earth.

—Michael Zeilik and John Gaustad, *Astronomy: The Cosmic Perspective* that opens the door just enough so you could look in to determine the status of the cat. You could peek in *but you decide not to*. Now what does QM predict for $\Psi$?

## 158. Wave Functions

Wave functions can be functions of many different physical parameters of the system of interest. For example, one can define a wave function in coordinate space, in momentum space, in spin space, and so on as long as the unit vectors of the space are orthogonal. For a single particle, the wave function $\psi(x_1, y_1, z_1)$ is the QM amplitude for finding the particle at the three-dimensional configuration space point $(x_1, y_1, z_1)$, which directly corresponds one-to-one to position space coordinates $x_1$, $y_1$, and $z_1$ for this one-particle system. For the two-particle system, the wave function $\psi(x_1, y_1, z_1; x_2, y_2, z_2)$ defines a six-dimensional configuration space. Is there a direct correspondence to three-dimensional position space coordinates for this two-particle wave function as well? What about the multiparticle wave function?

## 159. Wave Function Collapse?

Consider an electron in a box. Imagine partitioning the box into $N$ identical cubes and assume that the amplitude $\Psi$ for finding the electron in the box is the superposition $\Psi = \psi_1 + \psi_2 + \psi_3 + \ldots$, that is, the sum over all $N$ imagined identical cubes in the box. Now use a photon to observe where the electron is by recording the scattered photon and so on. Suppose your incident probe photon passes right through the box and does not interact with the electron, which you determine because the photon took a straight-line path to your detector. What happens to the electron wave function $\Psi$?

## 160. Quantum Computer

The new quantum computers rely on quantum coherence. That is, the quantum computer system contains $N$ identical quantum subsystems—for example, atoms, or optical setups, or molecules, or resonant cavities. In general, each quantum subsystem can be in many possible quantum states. Assume that the $\psi_i$ for each quantum subsystem has only two states, which we label 1 and 0. If $N = 3$, then $\Psi = \psi_1 + \psi_2 + \psi_3$ is the QM state of the system. Therefore our quantum computer represents all eight states simultaneously: 000, 001, 010, 011, 100, 101, 110, 111.

$$\Psi = {\downarrow}{\downarrow}{\downarrow} + {\downarrow}{\downarrow}{\uparrow} + {\downarrow}{\uparrow}{\downarrow} + {\downarrow}{\uparrow}{\uparrow} + {\uparrow}{\downarrow}{\downarrow} + {\uparrow}{\downarrow}{\uparrow} + {\uparrow}{\uparrow}{\downarrow} + {\uparrow}{\uparrow}{\uparrow}$$

That is, during calculations on $\Psi$ all eight states participate in each calculation! If the quantum computer is actually a large molecule in a vacuum, then the molecule must be kept away from the walls of the container and away from other molecules. Why?

## 161. Cup of Java Quantum Computer

One day while looking into her cup of java, Laura realized that this slurry of caffeine molecules could be the world's natural quantum computer. How could this inherent ability in coffee be possible?

## 162. Bragg Scattering of X-rays

Bragg scattering of X-rays of wavelength $\lambda$ in an ideal crystal satisfies Bragg's law: $2d \sin \theta = m \lambda$, where $d$ is the spacing between adjacent scattering planes and $\theta$ is the angle measured from the surface of the crystal, not

QUESTION: WHAT IS "IT"?

Pascal did IT under pressure.

Coulomb got all charged up about IT.

Hertz did IT frequently.

Boltzmann did IT in heat.

Ampere let IT flow.

Heisenberg was never sure whether he even did IT.

Bohr did IT in an excited state.

Pauli did IT but excluded his friends.

Hubble did IT in the dark.

Theorists do IT on paper.

Astrophysicists do IT with young starlets.

ANSWER: IT = science, of course!

—COPYRIGHT © 2002 BY JUPITER SCIENTIFIC

It has been proved that the 13th is more likely to fall on Friday than on any other day of the week. For a short proof consult the reference below.

—JOHN WAGNER AND ROBERT MCGINTY, "SUPERSTITIOUS?" *MATHEMATICS TEACHER* 65 (1972): 503–505 the perpendicular. When this condition is met for various integer values of $m$, constructive interference from the entire family of parallel planes occurs because the path differences are integral multiples of the X-ray wavelength. One often reads that the Bragg scattering of X-rays from an ideal crystal is a coherent scattering process—that is, all the Bragg-scattered X-rays arrive in phase at the detector. Why is it not so?

## 163. Beautiful Faces

Why can we see a person's face in great detail in visible light? Hint: think about coherent scattering versus noncoherent scattering of the light.

Why is the image of a person's face blurry in the infrared (IR) and in the ultraviolet (UV)? For simplicity and idealization purposes, assume that we can see equally well in the IR, visible, and UV so that our physiology is not the limiting factor.

## 164. Gravitational Waves

In addition to telescopes for photons in the $\gamma$-ray, X-ray, UV, visible, IR, $\mu$-wave, and radio parts of the electromagnetic spectrum, new windows to the universe are opening up with neutrino and gravitational wave observatories. Gravitational waves are expected to be produced by a changing mass quadrupole—for example, two masses revolving about their common barycenter, such as the two stars in a binary star system. They would emit gravitational waves with wavelengths of many kilometers that interact with all objects—that is, they exhibit most wave phenomena such as scattering, reflection, and transmission through objects in ways similar to other types of waves. The classical scattering cross section of gravitational waves by a mass pair in a detector was worked out by physicist J. Weber about 50 years ago.

For simplicity, assume that each pair of identical atoms in a material is a mass pair quadrupole scatterer of gravitational waves. We would like to know whether gravitational waves can scatter coherently in the detector—that is, whether a gravitational wave can simultaneously scatter from many mass pairs in the detector (such as an aluminum bar) or whether a gravitational wave must scatter from a single mass pair at a time. What is the physics here?

# 165. Coherent Neutrino Scattering

Another possible window or telescope for observing the universe is in the detection of neutrinos. The Super-Kamiokande neutrino facility in Japan and the Sudbury Neutrino Observatory (SNO) in Canada house two of the largest neutrino detectors, containing thousands of tons of water. Already they have determined that the flux of solar neutrinos from the Sun agrees with the standard solar model. In addition, research groups operating these neutrino detectors have verified neutrino oscillations in matter, the conversion of one type of neutrino to another.

The two neutrino detectors are enormous because neutrinos are notorious for their extremely small probability to interact with matter. Billions of neutrinos pass through our bodies each second and do no harm! A single electron neutrino would pass through solid lead (Pb), filling space from Earth to Jupiter with only a small chance of colliding with a Pb nucleus. However, in 1984 physicist J. Weber proposed that neutrinos of *all energies* could be coherently scattered by the nuclei in large defect-free single crystals of silicon, ruby, or diamond, thereby enhancing the neutrino scattering probability by a factor of $10^{22}$. Therefore, in the ideal case, practically all incident neutrinos would scatter at least once from the carbon nuclei in a perfect diamond crystal within the first centimeter or less!

Normally, one might expect only neutrinos of wavelengths much greater than the spacings between the nuclei in the crystal to have any chance at coherent scattering, analogous to light scattering coherently from a surface of atoms spaced much less than the wavelength of the incident light. Otherwise, when the nuclei are treated as scattering potentials, the phases contributed by the scattering nuclei to the QM amplitude are random, and the scattering probability will be proportional to $N$ instead of $N^2$, like the result for X-rays discussed in a previous problem. What assumption have we made about the scatterers that Weber says leads to an incorrect conceptual argument against coherent scattering for the shorter-wavelength neutrinos?

## 166. Magnetic Resonance Imaging (MRI)

Magnetic resonance imaging (MRI) is really the medical application of nuclear magnetic resonance, which physicists have been doing since the 1940s. A sample of living tissue contains numerous hydrogen atoms bound in molecules. Each hydrogen nucleus has a spin with a magnetic moment that can be aligned by an applied magnetic field. The sample is placed in a very strong uniform

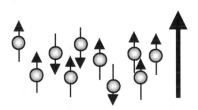

**Magnetic Field**

magnetic field to align the spins of the hydogen nuclei. A pulsed electromagnetic field is applied that would flip just one hydrogen spin, for example. What alternative QM interpretation can one provide that treats the nuclei as a collective whole?

# 167. Heisenberg Uncertainty

The Heisenberg uncertainty principle, also known as the indeterminancy principle worldwide, states $\Delta p_x \Delta x \geq h/4\pi$, where $\Delta x$ is the uncertainty in the x-position measurement, $\Delta p_x$ is the uncertainty in the x-momentum measurement, and h is Planck's constant. As some people say, the uncertainty principle places a limit on the accuracy of knowing a particle's position. What do you think? Some people claim also that the Heisenberg uncertainty principle is just an example of a more general uncertainty relationship for all waves, that the position can be determined only at the expense of our knowledge of its wavelength. Is this statement true?

We also know that Niels Bohr, in his discussions with Albert Einstein over several decades on whether quantum mechanics is a complete description of nature, would often invoke the uncertainty principle to defend his point of view, known as the Copenhagen interpretation of QM. Bohr argued that if you pin down the particle's position more precisely for the famous double-slit experiment by observing with photons, their interaction with the particle disturbs its momentum by giving it a random momentum kick. That is, without looking, the particle exhibits an interference pattern on a distant screen behind the two slits. However, if you look to see which way the particle goes through the slits, the measurement disturbs the system and there's no interference pattern on the distant screen. You just see a classical two-hump distribution. What do you think about Bohr's argument?

The number of photons is in general not conserved in particle reactions and decays. I ... would like to note here an ironical twist of history. The term "photon" first appeared in the title of a paper written in 1926. The title: "The conservation of photons." The author: the distinguished physical chemist Gilbert Newton Lewis (1875–1946) from Berkeley. The subject: a speculation that light consists of "a new kind of atom ... uncreatable and indestructible [for which] I ... propose the name photon." This idea was soon forgotten, but the new name almost immediately became part of the language.

—ABRAHAM PAIS, IN *SOME STRANGENESS IN THE PROPORTION*, EDITED BY HARRY WOOLF

## 168. Vacuum Energy?

Although the classical vacuum is a void, the quantum vacuum is a virtual "soup" of particle-antiparticle pairs that interact with real atoms to produce the Lamb shift (slight energy shift in atomic levels) and the Casimir effect (attraction of two plates in a vacuum). Does the quantum vacuum have energy content, or does the energy in the "soup" average out to zero?

## 169. Casimir Effect

When two parallel uncharged metal sheets are placed in a perfect vacuum, they attract each other with a tiny force that is not gravitational. What is the source of this effect?

## 170. Squeezing Light

Laser light can be described in many ways. If one considers just the amplitude and the phase of one ray in a laser beam, there will always be shot noise—that is, random variations caused by virtual particle interactions in the vacuum with the beam. Yet we've heard that there may be techniques to reduce the shot noise in the amplitude, for example. Then what happens to the shot noise in the phase?

## 171. Electron Spin

Does the vacuum affect the spin of a particle such as an electron?

## 172. Superconductivity

One quantum mechanical effect that shows itself on the macroscopic scale is superconductivity. Cooper-paired conduction electrons in superconductors have total spin zero, that is, their paired spins are opposite, even though their spatial separation can be enormous—centimeters to meters, for example—because they have opposite momenta. These pairs can act like bosons of spin zero, which obey Bose-Einstein statistics. Any number of bosons can be in the same quantum state, that is, have the same four-momentum (e.g., defined by the energy and three-momentum) and spin. Therefore, all bosons in the same collective superconducting state have exactly the same energy. Yet this boson collective state in a superconductor has a small energy width. Any thoughts about the cause of this energy width?

## 173. Superfluidity

He-4 below the lambda transition temperature 2.7 K can be analyzed as a two-fluid liquid composed of He atoms in the normal state and He atoms in the macroscopic superfluid state. Superfluidity is a property of He-4 in the liquid state because He-4 atoms obey Bose-Einstein statistics. Many He-4 atoms can be in the same macroscopic quantum state—that is, the same momentum states for these atoms moving in the superfluid. If so, then why can He-3 at low temperatures also become a superfluid?

## 174. Gap Jumping

Tiny detectors on a person's head have been used to sense tiny fluctuations in the brain's magnetic field. These SQUIDS, short for superconducting quantum interference devices, are the most sensitive of any kind and rely on the Josephson Effect, in which the Cooper pairs of electrons in a superconductor can sometimes jump a physical spatial gap in the material to another part of the superconductor. Manufactured SQUIDS

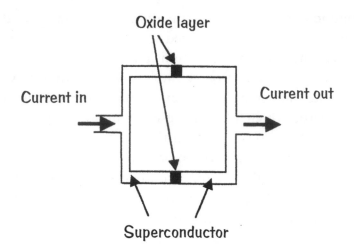

have a thin film filling the gap. In fact, the direct current (DC) SQUID used in laboratories worldwide today for sensing small magnetic fields is a superconducting ring with two gaps! The best DC SQUIDS have an energy sensitivity capable of detecting a magnetic flux change corresponding to about $10^{-34}$ joule in one second, about the mechanical energy required to raise an electron 10 centimeters in one second. Why do the paired electrons jump the gap?

# 175. Nuclear Decay

In the nucleus of an atom, neutrons and protons are held by nuclear forces. Their total energy (ignoring the $mc^2$ contributions) is less than the barrier height potential energy. Yet some nuclear particles do escape. Any thoughts about the reason for an escape?

# 176. Total Internal Reflection

In total internal reflection of light—for example, at a glass-air interface or from a water-air surface—if the incident light is in the more dense medium, does the light penetrate into the air beyond the interface?

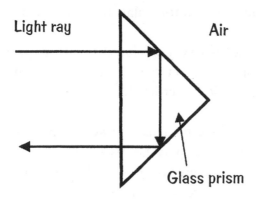

Light ray / Air / Glass prism

## 177. Annihilation

We know that particles and their antiparticles annihilate each other. For example, the electron and the positron in positronium can annihilate into two photons or three photons in the final state, depending on the total angular momentum of the positronium. Why would they do such a violent action?

Hint: why does any event occur in nature? We know that the rate of any quantum mechanical event, by Fermi's Golden Rule, is proportional to the probability for the event times the density of final states. Is this statement all we need to say?

## 178. A Bouncing Ball

We see a kid bouncing a ball. According to quantum mechanics, which applies to everything that happens, why does the ball bounce?

## 179. The EPR Paradox

First of all, a short explanation. Although there are other examples of the Einstein-Podolsky-Rosen (EPR) paradox and violations of Bell's inequalities, we choose this version because we can provide you with

The first wrist watches were decorative ornaments made for women, and men consequently shunned them as being too effeminate. Again war had its effect—the first World War. A watch that has to be fished out of a pocket in order to be read is a lot less convenient than one strapped to the wrist, particularly when you're trying to fire a machine gun or charge over a hill at a particular moment. Realizing this, governments made wrist watches part of the standard equipment issued to their soldiers. After the war, men suddenly felt that wrist watches were acceptable for civilian wear, a trend which clearly has continued for both sexes until the present day.

—Jo Ellen Barnett, Time's Pendulum: From Sundials to Atomic Clocks, the Fascinating History of Timekeeping and How Our Discoveries Changed the World actual data to use in formulating your own solution to the paradox.

A source of two correlated identical particles of opposite spins sits on the straight line between two identical particle detectors. Each detector can measure the polarization state of the entering particle, and each detector has three polarization switch positions (1, 2, and 3) and two display lamps (green and red). Each time the experimenter pushes the button, the two correlated particles are shot out of the source in opposite directions into the detectors. The data show two patterns: (1) For runs that have the same switch settings on the two detectors, the same color lights flash on them. (2) For all runs, without regard for switch settings, the pattern of flashing is completely random.

This experiment gets to the heart of QM and the application of its three rules for events. We can use classical mechanics to explain the first pattern: let the two particles carry the same instructions to be applied at the detectors. For example, this instruction set might work: flash red at switch positions 1 and 3; flash green at switch position 2. But this classical scheme with predetermined instruction sets will not handle the second pattern. Why not? What is the surprising conclusion?

Reproduced here is a small part of a data set for the experiment (from the Mermin reference in the answer). Each entry shows the switch settings and the colors the lights flashed for each run. The switch settings are randomly changed from run to run.

| | | | |
|---|---|---|---|
| 31GR | 13RG | 31RR | 33GG |
| 21RR | 31RG | 33GG | 11GG |
| 22RR | 12RG | 31RG | 13RG |
| 33GG | 13GR | 31RR | 31RG |

| | | | |
|---|---|---|---|
| 11GG | 22GG | 33RR | 23GR |
| 23RR | 12RG | 32RG | 31GR |
| 32GR | 12GR | 31RG | 23RG |
| 12GR | 22GG | 11RR | 22RR |
| 12RG | 23GR | 23GR | 12GR |
| 11GG | 33RR | 12GG | 32GR |
| 12GR | 23GG | 21GR | 12GG |
| 22RR | 23GG | 13GR | 31GG |
| 12GG | 33RR | 33GG | 32RG |
| 33RR | 23GR | 11GG | 21GR |
| 11RR | 21GG | 12RR | 22GG |

# 180. Information and a Black Hole

Classical information and quantum information are not the same. Why? Because QM rule 2 tells us that in QM there can be a coherent superposition of quantum states. No such state exists in classical physics. So quantum information supersedes classical information.

The classical and the quantum information content in a system, such as a chair, can be determined or estimated by standard techniques of classical and quantum information theory. Suppose the chair is tossed into a black hole. The quantum information in the chair seems to have gone with the chair into never-never land. Why should we worry about this information loss?

HEISENBERG'S UNCERTAINTY PRINCIPLE SIMPLIFIED

"If you know where it is, you don't know where it's going," or, "If you know where it's going, you don't know where it is."

The ordinary adult never gives a thought to space-time problems. . . . I, on the contrary, developed so slowly that I did not begin to wonder about space and time until I was an adult. I then delved more deeply into the problem than any other adult or child would have done.

—ALBERT EINSTEIN (TO NOBEL LAUREATE JAMES FRANCK) IN ALICE CALAPRICE, *THE EXPANDED QUOTABLE EINSTEIN*

# 9 | Can This Be Real?

A FTER QUANTUM MECHANICS EXPLAINED the internal behavior of the atom in the 1920s and the chemistry of atoms and molecules, physicists turned toward understanding the atomic nucleus in the 1930s and 1940s. Rutherford in 1911 had determined that practically all the atomic mass was in the nucleus, and of course everyone knew that its positive protons balanced the electron negative charges in the neutral atom. But what held the nucleus of positive protons together? A nuclear strong force was eventually identified in the 1970s as the color interaction acting between quarks, and it is one of the four known fundamental forces in nature. The second nuclear force, the weak interaction, responsible for many nuclear decays, was identified completely in the 1960s. By the early 1980s three of the four fundamental interactions had been unified into the Standard Model (SM) of Leptons and Quarks. Only gravitation needs to be incorporated into the unified model of nature. The selected challenges in this chapter range through the whole gamut of nuclear and particle physics.

# 181. Carbon-14 Dating

Carbon-14 is produced when cosmic rays collide with atoms in the atmosphere to create an energetic neutron that then collides with a nitrogen-14 atom (seven protons, seven neutrons) to make a carbon-14 atom (six protons, eight neutrons) and a hydrogen atom (one proton, zero neutrons). Carbon-14 is radioactive, with a half-life of 5,730 years.

These C-14 atoms combine with oxygen to form carbon dioxide, which plants absorb into plant cells through photosynthesis. Animals and people eat the plants and take in the C-14 as well as the normal non-radioactive isotope C-12. The ratio of C-14 to C-12 in the air and in all living things at any given time is assumed constant; about 1 in 10 trillion carbon atoms are C-14. The C-14 atoms are always decaying, so after an organism dies, no new carbon atoms are taken in and this ratio of C-14 to C-12 atoms decreases.

The carbon-14 radiocarbon dating of living and once-living materials began with Willard Libby in the 1940s. Antiquities dated by C-14 agree with other date records until they begin to disagree for dates more than several thousand years ago. Why is there disagreement in the dates between C-14 dating and the written records?

# 182. Nuclear Energy Levels

In the 1930s and 1940s, physicists working on the energy states of the nucleus of an atom concentrated on various models, including a shell model using the Schrödinger equation with an approximately constant electrical potential inside the nucleus. Conceptually, each nucleon is in a well-defined orbit within the nucleus and moves in an averaged field produced by all the other nucleons. However, even though quantum

Roughly once a second, a subatomic particle enters the earth's atmosphere carrying as much energy as a well-thrown rock. Somewhere in the universe, that fact implies, there are forces that can impart to a single proton 100 million times the energy achievable by the most powerful earthbound accelerators.

—JAMES W. CRONIN, THOMAS K. GAISSER, AND SIMON P. SWORDY, "COSMIC RAYS AT THE ENERGY FRONTIER," *SCIENTIFIC AMERICAN* (JANUARY 1997)

[Pierre Curie] was impressed by Marie's courage and her amazing love of work and fascinated by her lucidity, her challenging questions, her reflective answers.

—J. A. DEL REGATO, *RADIOLOGICAL PHYSICISTS*

On August 6, 1945, an atom bomb dubbed "Little Boy" was dropped from an American B-29 bomber called the *Enola Gay* on the city of Hiroshima. It detonated at 8:16 A.M. at a height of 1,900 feet. Of Hiroshima's 330,000 inhabitants, approximately 70,000 were killed instantly. By the end of 1945, the death toll had risen to 140,000. "Little Boy" used the gun assembly design and uranium-235 as the fissionable material. Because the gun design was an inefficient means of causing the chain reaction, about 50 kilograms of 89 percent U-235 and 14 kilograms of 50 percent U-235 ended up being used. Of this it is estimated that only about 2 percent actually fissioned. Three days later another atom bomb, dubbed "Fat Man," was dropped on the city of Nagasaki. Approximately 40,000 were killed instantly.

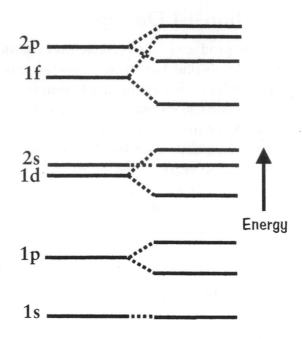

states such as $n = 1$, with $l = 0, 1, 2, 3$, etc., are possible in the shell model, the predicted energy levels did not fit the data. In fact, the actual energy levels were all scrambled compared to the shell-model theoretical predictions. Why?

## 183. Nuclear Synthesis

The championship of nuclear binding energy is often attributed to Fe-56, meaning that Fe-56 has the greatest binding energy per nucleon and therefore is the most stable nucleus. Most elements are synthesized in stars. Supposedly, elements higher on the periodic chart than Fe cannot be synthesized in normal star burning cycles. Why not? Actually, the sequence of nuclear synthesis does not stop at iron, because Ni also is synthesized. What happens to the Ni isotopes that are synthesized?

## 184. Heavy Element Synthesis

If we are "truly the stuff of stars," then where do all the heavier elements beyond iron come from if they are not made in normal star burning cycles?

## 185. Neutron Decay

A free neutron will decay with a half-life of about 14.8 minutes, but it is stable if combined into a nucleus. Why would the neutron be stable in the nucleus?

## 186. Finely Tuned Carbon?

Eventually a star exhausts its supply of hydrogen in its core, gravitational contraction occurs, the temperature reaches about $10^8$ K, and helium burning can occur via the reaction 3He-4 → C-12 + 2 photons. In fact, the nucleosynthesis of all the heavier elements essential for life relies on this reaction. However, the chance that three helium nuclei get together fast enough to form the carbon nucleus is negligible. So this critical reaction actually proceeds via an intermediate beryllium step given by 2He-4 + (99 ± 6) keV → Be-8 followed by

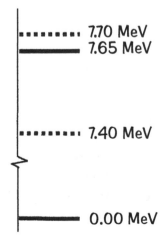

7.70 MeV
7.65 MeV

7.40 MeV

0.00 MeV

By the time she enrolled at the Sorbonne in 1891, Maria Sklodowska was twenty-four years old. In 1893 she passed the *license* in physics, coming first in her class. In that year Maria Sklodowska met Pierre Curie. The meeting between the two was arranged for scientific purposes, with little hint of matchmaking. At the time they met, both Maria and Pierre considered themselves destined for single lives. After graduation Maria intended to return to Warsaw to look after her aging father and teach science. Pierre, meanwhile, at age thirty-four one of France's leading young physicists, was convinced that he would never find a wife who would tolerate his complete devotion to science. He was the first to fall. Almost from the beginning he realized that in this severe Polish girl he had found the woman of his dreams.

—ADAPTED FROM MARGARET WERTHEIM, *PYTHAGORAS' TROUSERS: GOD, PHYSICS, AND THE GENDER WARS*; SUSAN QUINN, *MARIE CURIE: A LIFE*

Be-8 + He-4 → C-12 + 2 photons. Since the Be-8
lifetime of about $10^{-17}$ second is much longer than the
He-4 + He-4 collision time in a star, the beryllium will
be around long enough for the reaction to occur.

The total energy of the Be-8 nucleus and a He-4
nucleus at rest is 7.4 MeV above the energy of the normal state of the C-12 nucleus. The radioactive state of
the C-12 is 7.65 MeV above the normal state. If the
energy of the radioactive state were more than 7.7 MeV
above the normal state, the formation of C-12 via Be-8
plus He-4 would require the reactants to have at least
0.3 MeV of total kinetic energy, which is extremely
unlikely at the temperatures found in most stars.

The importance of this process is emphasized by
physicists who inject the Anthropic Principle, that certain constants of nature have values that seem to have
been mysteriously fine-tuned to just the values that
allow for the possibility of life. Recently, others have
introduced a further extension that claims that this carbon nucleus coincidence can be explained only by the
intervention of a designer with some special concern
for life. Both groups cite the closeness of the required
energy to the actual limit, 7.7 MeV – 7.65 MeV = 0.05
MeV, a quantity less than 1% of 7.65 MeV, as their evidence for the fine-tuning. Why is their reasoning sus-
pect with regard to this carbon formation process?

# 187. Proton-Proton Cycle

The thermonuclear reactions in the proton-proton
cycle inside the Sun convert four protons into an alpha
particle, two positrons, two electron neutrinos, and
two photons with the release of 26.7 MeV of energy.
First, two protons collide to form a deuteron H-2, then
this deuteron collides with a proton to form He-3, then
finally two He-3 nuclei must find each other to collide

and form an He-4. The overall representation of this proton-proton cycle is:

$$4H \rightarrow \text{He-4} + 2e^+ + 2\nu + 2\gamma.$$

The six photons ultimately produced, including the four 0.511 MeV photons from two positron-electron annihilations, take about a million years to reach the Sun's surface to be emitted eventually as visible photons, which then take about another eight minutes to reach Earth. The two neutrinos carry away about 3 percent of the energy to balance the energy conservation equation and to conserve lepton family number.

Presented as the primary source of our Sun's energy, this method of burning hydrogen is *not* the primary method for fusion energy in many stars. Why not? What reaction sequence is the primary candidate?

## 188. Oklo Nuclear Reactor

In the 1970s, uranium samples from the Oklo uranium mine in Gabon, Africa, were discovered to have abnormally high concentrations of the isotope U-235, as high as 3 percent, when only about 0.72 percent of the isotope was expected in a natural source. Supposedly the high concentration of U-235 is explained by realizing that the uranium deposits at Oklo acted as a natural nuclear reactor. Could this natural reactor have been a breeder reactor making its own Pu and U-235?

## 189. Human Radioactivity

Radiation doses are expressed in SI units as milliSievert (mSv) effective doses. This unit takes into account the type, the intensity and duration of radiation, the amount and type of body tissues irradiated, and the different radiation sensitivity of the irradiated tissues. The average natural background dose rate in many

Bertrand Russell would sometimes liken the scientific method to the following syllogism:

> Bread is made of rock;
> Rock tastes good;
> Therefore bread tastes good.

In other words, you can never be sure that correct conclusions don't follow from incorrect premises.

The illumination provided at eye level in artificially lighted rooms is commonly from 50 to 100 footcandles, or less than 10 percent of the light normally available outdoors in the shade of a tree on a sunny day. As a result, the total amount of light to which a resident of Boston, say, is exposed in a conventionally lighted indoor environment for 16 hours a day is considerably less than would impinge on him if he spent a single hour each day outdoors.

—RICHARD J. WURTMAN, "THE EFFECTS OF LIGHT ON THE HUMAN BODY," *SCIENTIFIC AMERICAN* (JULY 1975)

In early 1940, Paul Harteck, a German physical chemist, felt he'd need up to 300 kilograms of uranium to test his idea of using carbon dioxide as a moderator. He arranged to get the frozen carbon dioxide (dry ice) from I. G. Farben, and the necessary uranium from Heisenberg. But at the last moment, Farben declared they could only supply the dry ice until early June; they'd need it after that for keeping food fresh during the hot summer months. Harteck scraped together about 200 kilos of uranium, but with that low amount his results were inconclusive; Germany did not go ahead with the easy, dry ice reactor that would almost certainly have given them plenty of radioactive metal early on in the war. Thus was the clear hot weather of that summer—so often cursed by the Allies for letting Panzer armies advance into France—central to forestalling this greater evil.

—MARK WALKER, GERMAN NATIONAL SOCIALISM AND THE QUEST FOR NUCLEAR POWER 1939–1949

countries is 1–5 mSv a year. On average, medical exposures contribute about another 0.5–0.7 mSv a year. The current recommended limit for occupational exposure in many countries is about 20mSv effective dose per year averaged over five consecutive years.

The typical human adult body has an inherent internal radiation dose from its natural amounts of radioactive elements, including its major contribution of about 40 milligrams of radioactive potassium as the isotope K-40, which has a half-life of about 1.3 Gigayears. This isotope is not the result of artificial radioactivity but remains from the formation of potassium in the supernova that gave birth to our Solar System about 5 billion years ago. There has not been enough time for all of the radioactive potassium to decay, so that is why there is so much in our bodies. Eileen wonders whether this inherent K-40 radioactive source is exposing our bodies to more than the recommended limit? Is the limit exceeded when several people gather together in a small circle?

# 190. Nuclear Surprises?

Which of the following statements is true?

1. A typical coal burning power plant releases more radioactive materials into the air than a typical nuclear reactor plant.
2. Spreading all the nuclear waste equally around the surface of the planet will hardly change the background radiation level at all.

# 191. Cold Fusion

Is cold fusion—that is, the fusion of two deuterium nuclei at about room temperature—a possibility, or can this process be eliminated by theoretical arguments alone?

# 192. Fission of U-235

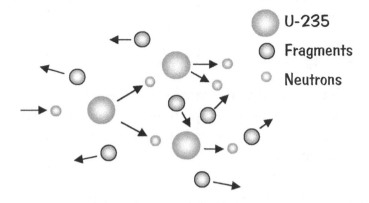

U-235
Fragments
Neutrons

During World War II the Germans and the Allies were both working on projects related to nuclear weapons development. One can calculate the minimum mass of U-235 required for a fission weapon from present-day nuclear physics data sheets. That value is the amount required if the neutrons produced by the fission of U-235 encounter stationary target nuclei. The problem is much more difficult for two important reasons. Can you identify them?

## 193. Minimal Nuclear Device

What is the minimum mass of pure U-235 or Pu-239 required in a device for a nuclear event? How would you estimate this value?

## 194. Large Nuclei

Small nuclei that become excited and deformed lose their energy by breaking up into smaller fragments. A larger nucleus, with 150 or more nucleons, stores most of its excitation energy as rotational energy. As they slow down and de-excite, these nuclei lose energy

Although we are quite unaware of their presence, there are, on the average, some 400 microwave photons in any cubic centimeter in the universe left over from the big bang.

In A.D. 499 the Indian astronomer Aryabhata presented a treatise on mathematics and astronomy, the *Aryabhatiya*. The *Aryabhatiya* is a summary of Hindu mathematics up to his time, including astronomy, spherical trigonometry, arithmetic, algebra, and plane trigonometry. The *Aryabhatiya* presented a new treatment of the position of the planets in space. It proposed that the apparent rotation of the heavens was due to the axial rotation of the Earth. Moreover, Aryabhata conceptualized the orbits of the planets as ellipses, a thousand years before Kepler.

—DICK TERESI, *LOST DISCOVERIES: THE ANCIENT ROOTS OF MODERN SCIENCE—FROM THE BABYLONIANS TO THE MAYA* and return to their unexcited shape. What do these nuclei emit, and how would you characterize the energy spectrum?

# 195. Human Hearing

The human eardrum is sensitive to displacements of less than the diameter of an atomic nucleus. How have such minute displacements been measured via nuclear physics techniques?

# 196. 1908 Siberia Meteorite

In an article by Andrew Chakin in *Sky & Telescope* in January 1984, pages 18–24, the author states:

> A grande dame of scientific mysteries—the Tunguska event—turned 75 last summer, her charm very much intact. She continues to seduce both scientist and charlatan alike, both hoping to explain what happened over a remote stretch of Siberian taiga on June 30, 1908. All that can be said from direct eyewitnesses is that a fireball nearly as bright as the Sun streaked to Earth out of a cloudless morning sky. The bolide's plunge was abruptly terminated by an explosion so great that it registered on seismic stations across Eurasia. The resulting shock wave circled the Earth twice.

The article relates that in 1908 in Siberia a huge meteorite is supposed to have crashed in the forest, causing huge fires and a crater many kilometers long, but no rocky debris was ever found.

By radiocarbon dating tree rings from old trees that have been living since 1908, Willard Libby and Edward Teller in 1963 may have learned something very important about the constitution of the meteorite. What could the radiocarbon data have suggested?

## 197. The Standard Model

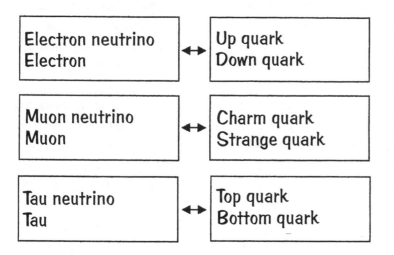

| | |
|---|---|
| Electron neutrino<br>Electron | Up quark<br>Down quark |
| Muon neutrino<br>Muon | Charm quark<br>Strange quark |
| Tau neutrino<br>Tau | Top quark<br>Bottom quark |

The Standard Model (SM) of Leptons and Quarks is the most successful physics model of all in terms of testing its concepts. The model has six leptons in pairs in three lepton families and six quarks in pairs in three quark families, with the quarks in three different colors. Aesthetically, the matching of three to three is pleasing. Mathematically, this matching of numbers of lepton and quark families cancels out infinities in quantum field theory calculations, such as the infinities that would arise from the famous triangle anomaly. However important this family matching may be, can you provide a fundamental physics argument for the specific matching of the first lepton family to the first quark family, of the second lepton family to the second quark family, and so on?

## 198. Spontaneous Symmetry Breaking

Spontaneous symmetry breaking is a concept first introduced by W. Heisenberg in describing ferromagnetic

The Schrödinger equation, published in March of 1926, was designed to explain almost all aspects of the behavior of electrons in terms of de Broglie waves, rather than of matrices. Physicists now could visualize the atom in terms of continuous processes—the ripple and flow of standing waves—whereas with matrices they had to deal with Heisenberg's assertion that the nature of the microworld was discontinuous and impossible to picture. Little wonder that many physicists threw away their matrices and started working with Schrödinger's methods. Even today, most physicists would say that the Schrödinger equation, being nonrelativistic, has no right to be this good.

—Adapted from Robert P. Crease and Charles C. Mann, *The Second Creation: Makers of the Revolution in 20th-Century Physics* materials. A ferromagnet has a perfect geometric symmetry until the Curie transition temperature is reached; then the material becomes magnetized and one particular direction of magnetization is chosen. The theory is symmetrical still, but the actual material is not. One can summarize the process by stating that microscopic events can have macroscopic consequences. Near the critical point of a phase transition, small, random fluctuations can grow to make their presence felt throughout the material. A few aligned spins can propagate their influence throughout the whole crystal, and the symmetry is broken.

Other examples are the Schrödinger equation and Maxwell's equations. As successful in helping to describe nature as they have been, these equations have more symmetry than the underlying phenomena they describe. Interest in their symmetry-breaking applications has led to significant new insights into new connections between macroscopic and microscopic phenomena.

In particle physics, the spontaneous symmetry breaking is achieved by the Higgs mechanism. The Standard Model of Leptons and Quarks relies upon the Higgs particle to spontaneously break symmetry to provide three of the electroweak bosons with mass while leaving the photon massless. Simultaneously, all the leptons and quarks get their mass values. Moreover, the effect of the Higgs field is to provide a frame of reference in the vacuum for the isotopic spin directions that distinguish the particles of each grouping—for example, neutrons from protons.

Is spontaneous symmetry breaking by the Higgs mechanism the only way to go? Are there other ways to spontaneously break symmetry to achieve the Standard Model of Leptons and Quarks?

## 199. Proton Mass

Kate sees that the chart of the fundamental leptons and quarks shows that the up and down quark masses are ~ 5 MeV/c$^2$ each. Yet the proton, which is composed of two up quarks and one down quark as the combination uud, has an enormous mass of 938 MeV/c$^2$. She asks why there is such a large mass difference between constituents and the final product.

## 200. Right- and Left-Handed Neutrinos?

Neutrinos are lepton family partners to the electron, muon, and tau particles of the Standard Model of Leptons and Quarks. Each neutrino is thought to be distinct, the electron neutrino being different from the muon neutrino, for example. We now know, however, that each lepton family neutrino type has a very small mass and is actually a linear combination of three fundamental neutrino states: $v_1$, $v_2$, and $v_3$.

For the weak interaction, there is the left-handed doublet state $|v_L, e_L>$ and the two right-handed singlet states $|v_R>$ and $|e_R>$, with the consequence that the right-handed states interact with the $Z^0$ boson but do not participate in the weak interaction mediated by the W$^+$ and W$^-$ bosons. The left-handed doublet interacts with all three weak bosons. Must one resort solely to the explanation "that is how Nature behaves," or is there another fundamental reason for left-handed doublet and right-handed singlet states?

## 201. Physics without Equations

John von Neumann and Stanislaw Ulam in the 1940s were among the first to consider attempting to

Contrary to the claim found in some dictionaries, the word *algebra* does not derive from an Arabic expression for *bone setting* but rather it means *compulsion,* as in compelling the unknown *x* to assume a numerical value.

The amount of ultraviolet radiation that penetrates the atmosphere varies markedly with the season: in the northern third of the U.S. the total amount of erythemal (skin-inflaming) radiation that reaches the ground in December is only about a fifteenth of the amount present in June.

—RICHARD J. WURTMAN, "THE EFFECTS OF LIGHT ON THE HUMAN BODY," SCIENTIFIC AMERICAN (JULY 1975)

In 1911 Marie Sklodowska-Curie, by then a double Nobel Prize winner, was asked to write a letter of recommendation for Albert Einstein who was being considered for a position at the ETH, the Zurich Polytechnic. She wrote, "In Brussels, where I attended a scientific conference in which M. Einstein also participated, I was able to admire the clarity of his intellect, the breadth of his information, and the profundity of his knowledge. Considering that M. Einstein is still very young, one is justified in placing great hopes in him and in regarding him as one of the leading theoreticians of the future."

—ALBRECHT FÖLSING, *ALBERT EINSTEIN: A BIOGRAPHY*

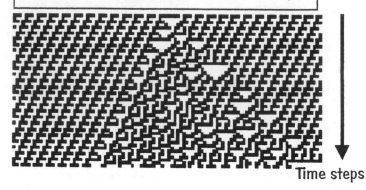

Cellular Automaton Rule 30 for 50 steps

Time steps

understand natural phenomena via cellular automata and computers. Cellular automata (CA) involve adjacent cells in a 1-D, 2-D, 3-D, and so on grid of cells (or nodes) that take on new numerical values at each tick of the clock according to given rules. The future state of each cell is determined only by the present state of its local neighborhood. One can even remove the external clock and still maintain a progression of states within the CA grid to simulate the passage of time.

Some people claim that all of nature will be simulated eventually on computers using cellular automata. Certainly, fluid flows and other large-scale systems in nature can be simulated to a reasonable degree by CA. But concerning the motion of electrons and other fundamental particles, which involves quantum mechanics and the fundamental interactions, how will these particles show their behavior with this CA technique?

# 10 | Over My Head

U NTIL THE 1920S NO ONE WAS SURE THAT WE were seeing stars outside our own Milky Way galaxy. Then, after Edwin Hubble established in 1927 that extragalactic galaxies existed and had recession velocities proportional to their distance from us, the cosmology game was afoot. The rules of the game had been established already by Einstein in 1916 with his general theory of relativity (GTR). The verification of one of its major predictions by analyzing the deflection of starlight passing near the Sun during the 1919 total solar eclipse told everyone that solid theoretical foundations were in place. But only in the 1990s did the vast accumulation of data on distant objects, by orbiting satellites such as the Hubble Space Telescope and the COBE microwave detector, and by a new generation of ground-based telescopes, transform a conjectural science into real testing of models of the universe. We present a sample of challenges from a vast range of possibilities.

The possibility that a massive object could bend light rays was discussed by Newton as early as 1704, and later by Henry Cavendish. However, the first actual calculation of the deflection angle was published by a Bavarian astronomer named Johann Georg von Soldner in 1803. Assuming that light was a corpuscle undergoing the same gravitational attraction as a material particle, Soldner determined how much bending would occur for a path that skimmed the surface of the Sun. The deviation, while small, is calculable, and Soldner's value was 0.875 seconds of arc. In 1911 Einstein, using the principle of equivalence, obtained the same result. Then in 1915, using the equations of the general theory of relativity, Einstein found that the deflection had to be 1.75 seconds of arc, twice the previous value. In recent decades the result as been confirmed to a precision better than 0.1 per cent.

—CLIFFORD M. WILL, *WAS EINSTEIN RIGHT?*

## 202. Olbers' Paradox

While walking through the fields one night with his dog, Jan looked up to see a remarkably clear night sky. In an instant, a famous question flashed in his mind: "Why is the sky dark at night?" With his engineering background, he determined that if the universe is uniformly filled with stars, then their successive spherical shells would contribute equal amounts, and the sky should be ablaze with light from all directions. Yet the night sky remains dark. What is the resolution of this paradox?

## 203. Headlight Effect

We live in a universe in which very distant stars have enormous cosmological redshifts of their light. This fact is interpreted as a cosmological recession velocity at nearly the speed of light. Unusual relativistic effects can be observed when looking at such fast-moving light sources. We consider a more local version here.

Suppose you are standing next to a straight test track that carries a vehicle with a light that shines in a cone with an apex angle 45 degrees about the forward direction. In the past, you have always seen the light as the vehicle approached. One day the vehicle for the first time is able to reach its highest speed ever: $v = 0.9999$ times the speed of light. But this time you do not see the light as it approaches. Why not? Do you see the vehicle as it passes and then recedes into the distance? Suppose a distant star or galaxy is approaching you at this speed. What would you see? And if receding?

## 204. Incommunicado?

In later problems we will encounter the behavior of light near a black hole, particularly its inability to escape from a black hole. That is, if you are trapped inside a black hole and are still alive, you cannot communicate with your friends outside because nothing escapes.

Meanwhile, consider a related problem in a normal space environment. Suppose you and your friend are in separate rocketships that begin next to each other and accelerate with respect to the stars in opposite directions. You both maintain steady pulsed light communication with each other via intense, nondiverging laser beams. But your relative speed is increasing each second as the separation distance grows ever faster. Will there come a time when neither of you will receive the other's light beam?

## 205. Local Accelerations

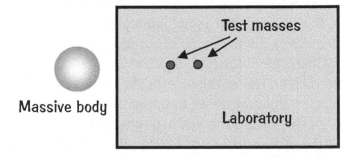

Einstein formulated the general theory of relativity (GTR) in 1915 based on his Equivalence Principle. In prerelativistic terms, a uniform gravitational field of strength g may be exactly simulated inside a rigid laboratory in a completely gravity-free region of space by accelerating this laboratory with a constant acceleration g m/s$^2$ relative to an inertial frame. By releasing two small test masses, their behavior reveals the physical environment.

Suppose an unseen massive body is near the rigid laboratory. What behaviors of the two small test masses will reveal its presence?

At age twelve Einstein suddenly became completely irreligious. Ironically, this conversion was the consequence of the only religious custom his parents observed, namely to host a poor Jewish student for a weekly meal. The beneficiary was Max Talmud (later "Talmey"), a medical student from Poland, ten years older than Albert. Talmud directed his attention to popular science books as well as to various books in mathematics. Einstein summed up the results of Talmey's influence: "Through the reading of popular scientific books I soon reached the conviction that much in the stories of the Bible could not be true. The consequence was a positively fanatic [orgy of] freethinking coupled with the impression that youth is intentionally being deceived by the state through lies. Suspicion against every kind of authority grew out of this experience."

—ADAPTED FROM MAX JAMMER, EINSTEIN AND RELIGION—PHYSICS AND THEOLOGY

Are the planets "arranged" so that the gravitational perturbations between them are smaller than would be expected from a random configuration, thus resulting in long-term stability of the system? Some argue that an example of this favorable arrangement is given by Neptune and Pluto, whose orbits appear to cross if we neglect their inclinations. Since their orbital periods are in the ratio 3:2 Pluto never gets near to Neptune and actually approaches more closely to Uranus. It is worth noting that the orbital planes of the planets are not coincident with each other or with the equatorial plane of the sun, which is inclined at about 7° to the ecliptic.

## 206. Twin Paradox

The twins are five years old when one of then is sent off in a spaceship that travels nearly the speed of light and the other remains on spaceship Earth. After 50 years Earth time the spaceship returns. The twins greet each other and compare their experiences. We know that the twin who experiences accelerations will age slower and return to Earth much younger than 55 years old. Precisely how does the general theory of relativity explain the aging of the twin during accelerations?

## 207. Twin Watches

This problem came from Richard P. Feynman in the 1960s while one of us (F. P.), an undergraduate, was with him in his car on the way to Malibu, California, where he gave weekly physics lectures. The third person in the car, B. Winstein, then a graduate student in physics, contributed to a discussion that became quite involved!

Charlotte holds two identical ideal watches at the same height, one in each hand. She holds one steady in her left hand and tosses the other into the air straight up. At the instant the upward-moving watch is alongside the other at the same height above the ground, she sees that the two watches are synchronized, with the exact same readout value. Later, on its downward free-fall path, she reads the time on both watches when they are again alongside each other and at the same height above the ground. Assuming that the moving watch is always in free fall, what would you predict for the two watch readings?

## 208. Global Positioning Satellites

The global positioning system (GPS) is a modern marvel, with a constellation of at least 24 satellites, each in a 12-hour orbit at an altitude of about 20,200 kilometers, whizzing around Earth at enormous speeds with respect to the ground beneath them. Each satellite knows its own position and sends out signals with this information. The GPS handheld receiver uses the signals from at least four different satellites to calculate its own position to within a few meters or better when a local reference signal is present. Yet within minutes the accuracy would reduce to many kilometers of error if one of Einstein's discoveries were not an essential part of the calculations in the GPS system. What are we referring to?

## 209. Solar Redshift

The light emitted from the Sun shows a redshift of the spectral lines even though our distance to the Sun is fixed during the measurement process. Why so?

## 210. Orbiting Bodies

When a body such as a planet orbits around a more massive body such as the Sun, the orbit does not close on itself, as expected from Newton's universal law of gravitation and Kepler's laws. The general theory of relativity (GTR) calculates the correct value for this precession of the orbital ellipse, determining that its complicated equations reduce to an equation similar in form to that of the classical Kepler problem, with an additional quadratic term that causes the precession. Can you provide a conceptual argument in GTR for the precession of the ellipse?

## 211. Gravitational Lensing

In examining the universe, astronomers utilize a technique called gravitational lensing of the light from distance stars. Supposedly, space itself can act as a lens for light rays. How can the emptiness of space—the vacuum itself—around stars and galaxies focus light?

The Russian astrophysicist George Gamow decided early in life that traditional religion could not be trusted. After watching Communion in the Russian Orthodox Church, he decided to see for himself whether red wine and bread could transform into the blood and flesh of Jesus. He held a bit of the blessed bread and wine in his mouth, ran home from church, and placed the specimen under the lens of his new toy microscope. It looked identical to an ordinary bread crumb that he had prepared at home earlier for comparison. "I think this was the experiment which made me a scientist," he recalled. In the 1940s, Gamow and others predicted the existence of a cosmic background radiation.

—COREY S. POWELL, *GOD IN THE EQUATION: HOW EINSTEIN BECAME THE PROPHET OF THE NEW RELIGIOUS ERA*

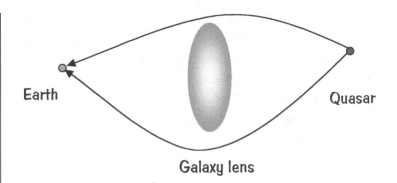

Earth        Quasar

Galaxy lens

## 212. Cosmological Redshifts

The light from a distant galaxy can exhibit a significant cosmological redshift. If the cosmological redshift is not a velocity redshift, what is its origin? Can the two effects be distinguished from each other by observing the spectrum of the galaxy or other light source?

## 213. Tired-Light Hypothesis

Since the 1920s there has been a popular hypothesis trying to explain the cosmological redshift as a so-called tired-light effect—that is, the light loses energy as its photons race through space, getting more tired with distance, like a long-distance runner completing a race. What two specific pieces of evidence rule out this explanation for the cosmological redshift?

## 214. Black Hole Entropy

A black hole has an entropy proportional to its surface area, so it must have a temperature above absolute zero. What would be evidence for this temperature?

## 215. Black Hole Collision

Two black holes collide head-on. Will they coalesce into one black hole?

## 216. Centrifugal Force Paradox

The general theory of relativity (GTR) predicts that in certain circumstances the centrifugal force may be directed toward, not away from, the center of circular motion. In fact, if an astronaut could steer a spacecraft sufficiently near to a black hole, the astronaut would feel a centrifugal force pushing inward, not outward, contrary to everyday experience! What is the conceptual explanation of the unusual result?

## 217. Geodesics and Light Rays

In conventional geometry, the geodesic is the shortest curve between two points measured by counting how many rulers fit along the curve. When learning relativity theory, one often reads statements that conflict with intuition, such as the following: "In any space-time, with or without a gravitational field, light always moves along geodesics and traces out the geometry of space-time." "In a space warped by a gravitational field, the light rays are curved and in general do not coincide with geodesics." Why are these phrases, taken from the general theory of relativity (GTR), not really in conflict with each other?

## 218. Galaxy Rotation

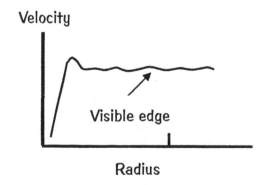

RIEMANN ANTICIPATES EINSTEIN

Ever since Newton, scientists had considered a force to be an instantaneous interaction between two distant bodies. However, over the centuries, critics argued that this action-at-a-distance was unnatural, because it meant that one body could change the direction of another without even touching it.

Georg Bernhard Riemann (1826–1866) developed a radically new physical picture, banishing the action-at-a-distance principle. To Riemann, *"force" was a consequence of geometry.* He concluded that electricity, magnetism, and gravity are caused by the crumpling of our three-dimensional universe in the unseen fourth dimension. Thus a "force" has no independent life of its own; it is only the apparent effect caused by the distortion of geometry.

—MICHIO KAKU, *HYPERSPACE: A SCIENTIFIC ODYSSEY THROUGH PARALLEL UNIVERSES, TIME WARPS, AND THE TENTH DIMENSION*

Venus always presents the same face to Earth when the two planets are at their closest approach, suggesting that its peculiar rotation may be due in part to terrestrial action. A simple calculation will show, however, that solar tidal action will dominate that of Earth on Venus, and so it is difficult to see how such a situation has evolved.

God is what mind becomes when it has passed beyond the scale of our comprehension. God may be either a world-soul or a collection of world souls. So I am thinking that atoms and humans and God may have minds that differ in degree but not in kind.

—FREEMAN DYSON

One of the great surprises in astronomy is the rotational behavior of galaxies—that is, all the stars in the galactic disk revolve at roughly the same tangential speed! Two immediate conflicts with conventional physics arise: (1) If Newtonian gravitation and Kepler's laws apply, they would dictate a decrease in star velocity with increasing radius from the galactic nucleus, like the planets of the Solar System do. (2) If the spiral arms of a spiral galaxy are to retain their integrity and persist for at least ten complete revolutions, as they have for the Milky Way, there must be something preventing them from wrapping numerous times. By assuming that Newton's universal law of gravitation applies to these galactic problems, what general type of mass/energy distribution must be proposed to explain the rotational velocity curve? What further hypotheses might you propose?

## 219. Cosmic Background Radiation

Cosmic background radiation was first detected in the microwave region in the 1960s and exhibits a perfect blackbody spectrum equivalent to the radiation from a source at a temperature of 2.72 K. One would expect

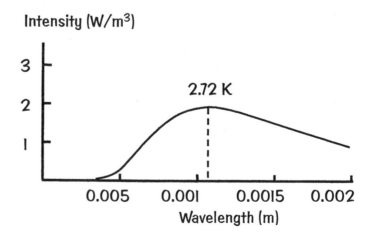

lots of remnant starlight all over the universe in all parts of the electromagnetic spectrum emitted for the past 10 billion years or so. Yet this microwave background radiation is not this light emitted from stars. How do we exclude this starlight?

## 220. Planetary Spacings

For some people the orbital radii for the planets in the Solar System seem to follow a regular pattern. The pattern was originally called the Titius-Bode law before Pluto was discovered. According to this numerical scheme, the semimajor axis of a planet's orbit $a = 0.4 + (0.3) 2^n$, where n is taken as negative infinity for Mercury, zero for Venus, and has increasing integer value by one unit for each successive planet. Neptune does not fit in this scheme, and the scheme may not represent an underlying physics.

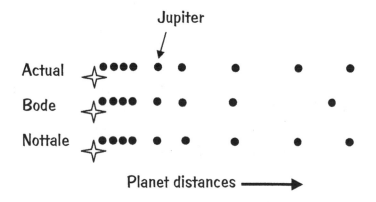

In the 1990s, applying chaos theory to gravitationally bound systems, L. Nottale found that statistical fits indicate that the planet orbital distances, including that of Pluto, and the major satellites of the Jovian planets, follow a numerical scheme with their orbital radii proportional to the squares of integers $n^2$ extremely well! The planets were divided into two groups, the inner

NOT ALL STARS ARE ROUND!

Many red giant stars, including the bright star Betelgeuse and the well-known variable star Mira, exhibit peculiar egglike shapes, presumably because of the huge convection currents roiling their filmy outer layers. Astronomers also found huge cocoons of hydrogen gas surrounding hot blue stars and clouds of titanium oxide billowing off red giants' surfaces.

—"MOONBALL: ASTRONOMERS BEAT A PATH TO HIGH RESOLUTION," SCIENTIFIC AMERICAN (JULY 1993)

Many of the oldest stars and star clusters in the galactic halo of the Milky Way move in retrograde orbits—that is, they revolve around the galactic center in a direction opposite to that of most other stars.

—SIDNEY VAN DEN BERGH AND JAMES E. HESSER, "HOW THE MILKY WAY FORMED," SCIENTIFIC AMERICAN (JANUARY 1993)

planets Mercury, Venus, Earth, and Mars being at $n = 3$, 4, 5, and 6, respectively, and the outer planets, starting with Jupiter at $n = 1$. The two sets can be combined into one set with Mercury at $n = 3$, Jupiter at $n = 10$, and so on. The lack of planets at some integers can be attributed to the history of the Solar System and does not indicate a failure of the prediction.

Other researchers claim that the Nottale sets of integers are not unique and that several alternative sets of integers exhibit excellent fits, raising the question of whether there is actually a unique pattern in the orbital spacings. In addition, there are known orbital resonances for the satellites of the Jovian planets that cause some of the apparent patterns in the satellite spacings.

How would you determine whether any of the claimed patterns are physically significant or simply numerology?

## 221. Entropy in the Big Bang

"The primordial fireball was a *thermal* state—a hot gas in expanding thermal equilibrium. But the term 'thermal equilibrium' refers to a state of *maximum* entropy. However, the second law demands that in its initial state, the entropy of our universe was at some sort of *minimum,* not a maximum!" How would you resolve this paradox raised by R. Penrose?

## 222. Gravitational Wave Detectors

Radio wave detectors are calibrated by sending out radio waves from a transmitter several wavelengths away and more. Why can't builders of gravitational wave detectors do the same thing? After all, one could put two large masses at opposite sides of a rotating platform and spin them around to have a gravitational wave source for detector calibration.

## 223. Space Curvature

The general theory of relativity (GTR) has been checked and verified at local distance scales. We know that the GTR may not explain the rotation curves of galaxies without the introduction of "dark matter." We do not expect the GTR to work for extremely small distances, for extremely short time intervals, or for cosmological distances—that is, whether the GTR correctly explains the universe on a global scale. The GTR, for example, allows for the overall curvature of space but does not predict its global value. In better words, the GTR does not fully predict the geometry of space, neither determining the global shape nor the connectedness of space.

Suppose you were given the task of measuring the overall curvature of space. One way might be to count the number of stars at each radial distance, say, and plot the number found versus distance. How does this method determine the curvature of space? Does this technique work for both continuous and discrete spaces?

## 224. The Total Energy

The total energy in the observable universe can be shown to be zero by adding the total mass energy in matter and radiation to the total gravitational potential energy. That is: energy total = mass energy + gravitational energy. Does this result mean that the creation of matter out of nothing contradicts no physical conservation law?

## 225. Different Universes?

The present limited understanding of our universe allows for much speculation with regard to whether we live in but one of many universes, possibly with

WHAT'S IN A NAME?
Planck happened upon *"Relativtheorie"* when in 1906 he was groping for a name to distinguish the Lorentz-Einstein theory (of the deformable electron) from Abraham's (theory of the rigid, spherical electron). It was in the discussion following Planck's 1906 lecture on Kaufmann's experimental electron data and their theoretical interpretation that Planck's term *Relativtheorie* was embellished by the experimentalist A. H. Bucherer to *Relativitätstheorie.* Although Einstein had in his published work from the very first referred to the *Relativitätsprinzip,* he did not use the designation *Relativitätstheorie* in print until 1907, as a variant of Planck's expression. Felix Klein's suggestion of *Invariantentheorie* did not catch on.

—ERIC SHELDON, "RELATIVITY OR INVARIANCE?," *AMERICAN JOURNAL OF PHYSICS* (SEPTEMBER 1986)

connections among them. These wild conjectures are permitted simply because we do not know enough about the origins of the fundamental constants such as Planck's constant, the gravitational constant, or the lepton and quark masses, for example. Indeed, the other possible universes could have different values for these constants. Suppose that the lepton and quark mass values are discovered to be determined by some fundamental properties in mathematics such as the invariants of elliptic functions. How might this discovery end speculation about many universes?

# 11 | Crystal Blue Persuasion

S OME PROBLEMS THAT COULD HAVE BEEN PUT in the previous chapters are presented in this chapter. We have collected these special problems for this grand finale. Some of the previous problems will provide significant clues toward answering challenges in this hodgepodge of a collection, while many challenges here are new to most readers. We hope that you enjoy them as much as we have.

## 226. Iodine Prophylaxis

Supposedly, in the event of a nuclear emergency, iodine tablets offer protection from radioactive iodine. How can this preventive measure work? Isn't all the iodine, tablets or not, exposed to the ambient radiation?

## 227. Bicycle Tracks

If you came upon this set of bicycle tracks meandering around in the mud, could you determine which way the bicycle was going simply by examining the tracks? Remember that the 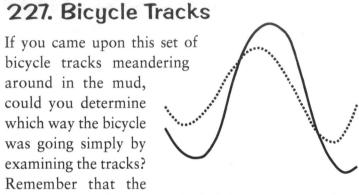 front wheel and the rear wheel make separate tracks.

## 228. Earth Warming

Over the past several decades there has been considerable concern over the possible slow rise of a few tenths of a degree Celsius in the average temperature of Earth's atmosphere and its surface. Most of the concern seems to be associated with the greenhouse effect on the radiation from the Sun. However, assuming that the average temperature of Earth can be defined unambiguously so that the small rise in average temperature is true, can there have been additional thermal energy coming from within Earth to cause this rise?

## 229. Frequency Jamming

Suppose one desired a noisy emitter of electromagnetic waves at all frequencies simultaneously. Such a device might be useful to jam undesired cell phone signals, for example. How could you do this simply?

## 230. Light Energy

We know that the speed of light is the same for all observers in inertial frames. If so, are the momentum and energy of light as measured by all observers the same even when the light source is moving toward the observer?

## 231. Acid Rain

Chemists define the quanity pH = -Log [$H^+$], where [$H^+$] is the aqueous concentration of $H^+$ ions. Pure rainwater is a neutral solution that has a pH of 7 when the droplets form. Will these droplets falling through clean, unpolluted air have a pH of 7 by the time they hit the ground?

## 232. Electrical Current

Upon turning on a lamp, Raymond wonders, "About how fast do the electrons in house wiring move as they provide electrical energy to the lamps and other electronic devices?"

## 233. Earth's Orbit

Earth's elliptical orbit is not fixed in orientation with respect to the stars. Why not?

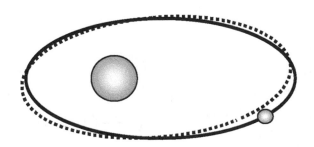

If the universe of science is not evident to our ordinary senses but is elaborated from certain key perceptions, it is equally the case that these perceptions require their appropriate instruments: microscopes, Palomar telescopes, cloud chambers, and the like. Again, is there any reason why the same should not hold for religion? A few words by that late, shrewd lay theologian, Aldous Huxley, make the point well. "It is a fact, confirmed and reconfirmed by two or three thousand years of religious history," he wrote, "that Ultimate Reality is not clearly and immediately apprehended except by those who have made themselves loving, pure in heart, and poor in spirit."

—HUSTON SMITH, BEYOND THE POST-MODERN MIND

I swear to you that to think too much is a disease, a real actual disease.

—FEODOR DOSTOEVSKY, NOTES FROM UNDERGROUND

## 234. Crystal Growth

Many children grow crystals in solution as a science project. How does a crystal grow from a small "seed" to its final size? That is, exactly how do the atoms know where to adhere to the growing structure without fouling up the precise cubic crystal structure development, for example? Do you see the dilemma? And the surprise?

## 235. Ruby, Sapphire, and Emerald

How are ruby, sapphire, and emerald crystals related? How do they produce their colors?

## 236. Kordylewski Clouds

By Kepler's laws any object orbiting the Sun in an orbit smaller than Earth's has a faster speed. So how can dust particles placed in solar orbit along the Earth-Sun radial line but closer to the Sun have the same speed as Earth?

## 237. Twist Scooter

There is a type of three-wheeled scooter with a handlebar extending vertically upward from the apex of a V-shaped horizontal metal frame that has three wheels. The front wheel can rotate about a vertical axis at the foot of the vertical handlebar at the apex

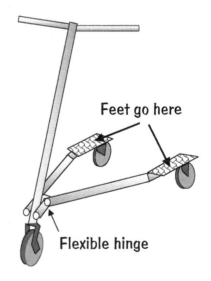

Feet go here

Flexible hinge

intersection, and the two rear wheels are at the ends of the arms of the V. The two arms of the V are hinged at their intersection at the front of the scooter so they can rotate about the vertical axis of the hinge—that is, they can form a wider or a narrower V angle within limits. The rider places one foot on each arm of the V and swivels his or her body from side to side to tilt the vertical handlebar from side to side to make the vehicle go forward or backward. Can you explain the physics of its forward motion?

## 238. Unruh Radiation

Physicist J. Bekenstein determined that a particle accelerating in a vacuum experiences a blackbody radiation bath around itself at a temperature directly proportional to its acceleration. By the equivalence principle, would a particle at rest in a gravitational field also experience this blackbody radiation bath?

## 239. Star Diameters

One can determine the diameter of a distant star even though the diameter cannot be measured by parallax. The process uses the *intensity* interference, not the

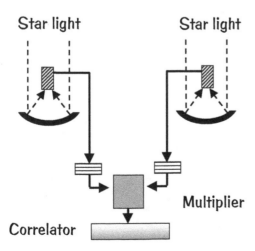

*amplitude* interference, between the light entering two identical photodetectors (telescopes) from the left side of the star surface, and from the right side of the star surface. This process is valid even though the star is simply a point in the light-gathering optics of either photodetector. Can you explain the physics?

## 240. Glauber Effect

Does a standard incandescent lightbulb emit single photons? Paired photons? Triplets?

## 241. Bird Sounds

Practically all birds and other animals emit sounds that have a fundamental frequency and several harmonics. Some birds, however, can emit just the fundamental with no harmonics! Measurements inside the bird reveal that the original sound has harmonics. So how does the bird eliminate them before they escape into the great outdoors?

## 242. Spouting Alligator

Some alligators can submerge themselves slightly underwater and vibrate their bodies so that numerous

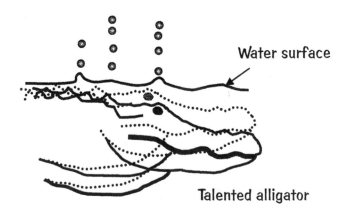

Water surface

Talented alligator

individual water droplets are projected upward simultaneously a foot or more directly above the backs of their heads. What is the physics here?

## 243. Hair-Raiser Function

One of us (F. P.) first heard about the hair-raiser function (HRF) from physicist Richard Feynman. We introduce this function here as a curiosity to stimulate the mind. And if one pulls a hair on top of the head upright, this function is probably a good representation of its fast vertical increase with horizontal distance.

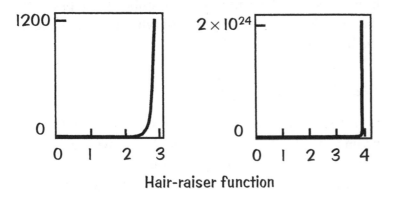

Hair-raiser function

Most mathematical functions are easy to define, and so are their inverses. Physicists utilize a tremendous variety of mathematical functions, the two most common being the power of a quantity and the exponential of a quantity. Physicists also use mathematical operations that may not seem to qualify as a function, such as the Dirac delta function $\delta(r - r_0)$. Powers are prevalent in fundamental laws dictated by geometrical symmetries such as the universal law of gravitation and the Coulomb law, both having potentials proportional to $1/r$ and forces proportional to $1/r^2$ for ideal point sources. The exponential function increases more

When Glenn T. Seaborg and his colleagues at the Lawrence Berkeley National Laboratory in California were able to make a new element in 1940 with 94 protons in its huge nucleus, they could not at first imagine that anything more massive would ever be obtained, and so they called their new element *ultimium* (later it would be renamed plutonium).

A typical person age fifteen to forty-five grows about an inch overnight. When we lie down to go to sleep at night, the spaces in our spine expand and we get taller. In the morning when we get up, gravity reasserts its downward pull and very quickly we go down a bit. Astronauts typically stretch more than two inches in weightlessness, but as soon as they return to Earth gravity, over a matter of hours they shrink to normal height.

Cosmic radiation comes from outer space. The radiation dose from cosmic radiation increases with altitude, roughly doubling every 6,000 feet. Therefore, a resident of Florida (at sea level) on average receives about 26 millirem, one-half the dose from cosmic radiation as that received by a resident of Denver, Colorado, and about one-fifth of that by a resident of Leadville, Colorado (about two miles above sea level). A passenger in a jetliner traveling at 37,000 feet would receive about 60 times as much dose from cosmic radiation as would a person standing at sea level for the same length of time.

I don't think there is one unique real universe. . . . Even the laws of physics themselves may be somewhat observer-dependent.

—STEPHEN HAWKING

rapidly than practically any other known function and is used whenever the change in a quantity is proportional to the quantity itself, such as in growths and decays.

The hair-raiser function HRF($x$) can be defined by example on how it maps integers to integers. The HRF(1) is 1. The HRF(2) = $2^2$. The HRF(3) = $(3^3)^3$, and so on. Notice the grouping with parentheses. One needs a calculator for most of the higher-integer HRFs. Certainly the HRF is a one-to-one mapping.

How does one calculate the HRF of a non-integer? Of a complex number? How does one determine the inverse of the HRF? That is, given a number such as 42, how does one determine what number is mapped by the HRF into 42? And finally, of what potential use is the hair-raiser function?

## 244. Space Crawler

In 1999, the U.S. Patent Office awarded a patent to a propulsion device that is a base frame with a sliding carriage on the frame and two counterrotating masses that together couple and decouple to the base frame to move the carriage forward and backward in a complicated motion. An onboard battery provides the energy for all internal movement. When the rotating masses are not coupled to the frame, and when placed on a nearly frictionless air table, the whole device simply oscillates forward and back repeatedly, as expected. When the coupling is allowed to occur at specific phases in their rotational cycle, the whole device moves only forward in a continual sequence of spurts that are longer with less air table friction! Will this device operate likewise in space?

# Answers

## Chapter 1
## The Heat Is On

### 1. Egg into a Bottle

Newton's second law explains the result. When the egg is resting on the bottle, the ambient air in the room plus the gravitational force on the egg by the Earth together exert a total downward force on the egg that is equal to the total upward force provided by the contact force of the bottle plus the force of the air inside. By Newton's second law, a downward acceleration begins when there is a net force downward. To get the egg to accelerate downward, you must reduce the upward force of the air inside the bottle. This action reduces the total upward force, allowing a net force downward, with the resulting acceleration downward dictated by net force = mass times acceleration.

The correct timing requires one to wait until the paper dropped inside the bottle has stopped burning, then immediately and carefully place the egg on the opening. The warmed air inside the bottle will begin to cool just as the burning finishes. The egg seals the opening, so the air pressure inside decreases as the cooling progresses. The total downward force will eventually be greater than the total of the forces resisting the egg at the entrance, and the net force downward will accelerate the egg into the bottle. The movement continues until the egg drops to the bottom. Kerplop!

### 2. Egg out of a Bottle

The hard-boiled egg is in the bottle and the goal is to remove this egg without damage. When the total outward force acting on the egg exceeds the total opposing force resisting its exit, then the egg will accelerate out of the bottle. Assume that the bottle is held vertically upside down. Earth's gravitational force acting on the egg—that is, its weight—and the force of the air inside the bottle both contribute to the total downward force acting on the egg. The upward force is the contact force of the bottle (which includes static friction) plus the force of the ambient air in the room.

To remove the egg, you need to create a pressure difference between the air inside and the air outside the bottle, with the greater pressure inside. Hold the bottle with its mouth near your mouth and its bottom somewhat higher so that the egg is positioned far forward, to the neck of the bottle, but not completely blocking the neck. Exhale a tremendous burst of breath into the bottle to suddenly increase the inside air pressure. The Bernoulli effect, the reduction in air pressure perpendicular to the air stream flow direction, caused by the rush of air flowing around the egg plus the increased inside air pressure, will aid in pushing the egg through the neck and mouth of the bottle. When the net force outward occurs, the egg accelerates outward. Sometimes the egg just pops out and must be caught in the mouth, and at other times the egg must be gently removed at the opening. The more vertical the bottle is, the more help one has from the gravitational force of the Earth.

# 3. Sugar

One can usually dissolve about five cups of sugar in one cup of water! Very simply, sugar molecules can squeeze into the empty spaces among the water molecules, so they are not really occupying more space. The water forms somewhat of an open latticework with water molecules loosely connected, so the "holes" in this water lattice can accommodate a large number of other molecules. The sugar molecules form temporary hydrogen bonds to the water molecules, these bonds breaking and re-forming continually. Essentially the rule is "like dissolves like." Of course, the sugar molecules are quite large, and one cup of sugar molecules contains only about one twenty-fifth the number of water molecules in a cup. So there are many water molecules to every sugar molecule in the solution.

Wolke, R. L. What Einstein Told His Cook: Kitchen Science Explained. *New York: W. W. Norton, 2002, pp. 21–22.*

# 4. Kneading Bread

Each successive kneading of the bread dough distributes the $CO_2$ gas released by the action of the yeast to make a finer texture—that is, smaller holes more evenly distributed throughout the bread volume.

Initially, the concentration of yeast is not uniform but has some non-uniform volume distribution in the bread dough. Where there is more yeast, there will be more $CO_2$ gas produced by the yeast chemistry and usually bigger bubbles in the region. At the molecular level, $CO_2$ molecules released by the yeast will diffuse into the surrounding dough somewhat, probably not moving very far in the available time. Some of the gas bubbles may even coalesce to form bigger bubbles. Without further kneading, some places in the bread will possess many bubbles or large bubbles, and other places may have very tiny bubbles or none. We have all seen bread with a non-uniform distribution of bubbles, or even with one large bubble somewhere. A more thorough kneading would eliminate these oddities, unless they are planned.

# 5. Measuring Out Butter

Butter floats on water because its density is less than the density of water. The recommended procedure in many cookbooks for measuring one-half cup of butter is: put one-half cup of water into the measuring cup and then add pieces of butter until the water level is pushed up to the one-cup level. Often a reference to Archimedes' principle is given. (According to Archimedes' principle, a body wholly or partially immersed in a fluid will be buoyed up by a force equal to the weight of the fluid that it displaces.)

However, this recommended measuring procedure does not use Archimedes' principle! And the measured amount of butter is not exactly one-half cup!

If ice chunks were used instead of butter, then the procedure is correct. As a check, when floating ice melts, the water level does not change. Therefore, the recommended procedure is correct for measuring one-half cup of ice. But the density of butter is not the same as that of the ice, and the density of melted butter is not the same as that of the water. Therefore, the immersed volumes of ice and butter would be different.

However, if the butter is held under the water surface when water is added, then the butter measurement is correct when the water reaches the one-cup level. One is simply measuring the volume of the butter, and one does not use Archimedes' principle for this measurement.

# 6. Milk and Cream

The milk is "heavier," meaning more dense. Cream floats to the top in a milk-cream mixture, so the cream is less dense. Quite often people confuse mass density with liquid flow sluggishness. The two properties are unrelated. Many of these people will think that the cream is more dense. Anyone who has milked a cow or a goat has seen the cream at the top. A pint of light cream weighs about 1.5 grams more than a pint of heavy cream on average. Why? Because heavy cream has more fat per volume, and fat is less dense than water.

Separation of materials by density has been an important method for millennia. Gold and platinum have been separated from other elements by dumping the ores into a hot bath of lead. The specific density of lead is 11.36, of gold 19.32, and of platinum 21.45, so gold and platinum atoms sink and practically all other elements and compounds float. Of course, the workers must not breathe in the lead atoms in the vapors.

# 7. Straw and Potato

Simply placing the straw in one hand and trying to push the straw through the hard potato leads to a frustrating failure. The straw material—paper or plastic—cannot take much compression before bending sideways. So this sideways bending must be prohibited if success is possible. We can use air pressure to help make the straw more rigid.

Pinch the straw between your thumb and forefinger about two inches from the end that is farthest from the potato and squeeze tightly. Hold the potato carefully but securely in the other hand placed in a horizontal plane with thumb on one side and fingers on the other side of the potato. Make sure that no part of the hand will be in the path of the straw—that is, avoid having any part of your hand on the top or the bottom of the potato. With a sudden thrust, drive the pinched straw into the potato held in the other hand. The straw goes right through. Why? The trapped air upon contact is compressed inside the straw and helps the straw remain rigid. The paper or plastic is stretched taut and is more difficult to bend significantly. Alternately, one could pinch the straw in a vertical position in a vise or a clamp and drop the potato onto the straw (or a collection of straws).

# 8. Blueberry Muffins

The downward drift of the blueberries in the warm batter is caused by the gravitational force of the Earth, so one must increase the friction between the batter and the surface of the berry to hinder this settling. One could make the batter thicker, but this solution may not produce the desired muffin texture. Instead, before mixing the blueberries in the batter, dampen them slightly and shake them in a bag of flour. The flour attaches to the berry surface and increases the static friction with the batter, keeping the berries uniformly distributed in the muffin.

The physics involves Newton's second law, found in every high school textbook, that is, net force = mass × acceleration. The downward gravitational force on the blueberry must be balanced by an upward force applied by the batter through friction to produce zero net force in the vertical direction and therefore zero acceleration downward from rest.

In this case, the maximum value of the static frictional force (equal to the coefficient of friction times the force of the batter *perpendicular* to the blueberry's attempted movement) has not been exceeded. Without the flour coating, the static friction coefficient is too small. The maximum static frictional force upward provided by the batter alone is too small. The berry accelerates downward until reaching the critical velocity, for which the net force is again zero, so the blueberry drifts downward with no acceleration. With the flour coating, the coefficient of static friction is large enough to not be exceeded and the downward gravitational force is always balanced by the static frictional force upward.

At the atomic level, friction involves electrical forces acting between atoms and between molecules. Still an active research area, the influence of quantum mechanical effects is significant. Even sound waves contribute good vibrations in this field, called tribology!

Krim, J. *"Friction at the Atomic Scale."* Scientific American 275, no. 4 (1996): 74–80.

Miller, J. S. The Kitchen Professor. *Sydney: Australian Broadcasting Commission, 1972, pp. 22–24.*

# 9. Can of Soup

Turn the can of soup upside down and open the bottom. Then turn the can over and watch the concentrate being pushed out by the weight of the more-liquid stuff, assuming that the upward-acting static frictional force of the wall with the concentrate balances the weight of the concentrate. The weight of the liquid thus provides the nonzero net force downward to accelerate the liquid downward by Newton's second law.

If there is delay in this evacuation process, allowing some air into the liquid region behind the solid concentrate might expedite the motion. Sometimes the air seal at the wall is very good, so that as the concentrate slips outward, a significant inward pressure difference can build up to slow down the extraction process. In addition, the molecules in the soup may interact more vigorously than expected via the electrical force between the concentrate and the wall of the can (i.e., the viscous force, the surface tension, etc., may be large enough to make the extraction even more challenging).

# 10. Salt and Sugar

Salt and sugar do their work on bacteria by osmosis, dehydrating them so that they die or are deactivated. A bacterium in very salty water has a saltier environment outside its cell membrane than inside. Water molecules will move from its inside through its water-permeable membrane to the outside to balance the salt concentrations, a process called osmosis. The bacterium shrivels up and dies. Sugar works by the same process to preserve fruits and berries. In the markets today one can buy "cured" hams and other pork products that use both salt and sugar to enhance the flavor.

*Wolke, R. L.* What Einstein Told His Cook: Kitchen Science Explained. *New York:* W. W. Norton, *2002, pp. 137–138.*

## II. Defrosting Tray

The "miracle" defrosting tray is simply aluminum metal, and one could use a thick aluminum frying pan or other piece of aluminum or copper metal tray to do the thawing as quickly, as long as there is no coating on the metal. Nonstick pans have coatings that are poor heat conductors. Metals are the best heat conductors because they have about $10^{23}$ conduction electrons per cubic centimeter available to transfer thermal energy from the hotter source to cooler regions. For thawing purposes, the metal tray will conduct thermal energy from the room air into the frozen food very efficiently.

*Wolke, R. L.* What Einstein Told His Cook: Kitchen Science Explained. *New York:* W. W. Norton, *2002, pp. 202–203.*

## 12. Ice Cream Delight

Good ice cream contains abundant air bubbles to keep it light with very small ice crystals so that the texture is smooth. There are several good small appliances for making ice cream and sorbet. Some ice cream makers rely on human musclepower to turn a large spatula to control the ice cream texture; others are electric. But with an abundant supply of liquid nitrogen at −196°C and about an equal volume of ice cream mixture, the freezing of the ice cream occurs so fast that only small crystals have time to grow. As the nitrogen furiously boils, plenty of small gas bubbles form in the mixture. All these effects make for a delicious treat.

*Kurti, N., and H. This-Benckhard. "Chemistry and Physics in the Kitchen."* Scientific American 270, no. 4 (1994): 66–71.

*Walker, J. "The Physics of Grandmother's Peerless Homemade Ice Cream."* Scientific American 250, no. 4 (1984): 150–153.

## 13. Cooking a Roast

The bone-in roast cooks faster because the bone, even though porous, rapidly conducts thermal energy to the inside faster than the meat itself does. So the bone-in roast cooks from both directions—outside in and inside out. There will be some minor speedup effect from the difference in specific heats of bone and meat, and there will be slightly less meat if both roasts weigh the same, but in the simplified, ideal case we ignore these differences. One could use computer modeling with the appropriate physics equations to determine the temperature distribution in various parts of the meat and bone in the two cases, showing that some parts of the meat are cooked more than others in both types of pieces.

Any general physics text discussing thermal conductivity and specific heat contains the pertinent information for analyzing this problem, but the actual temperature profile as a function of the elapsed time is difficult without some idealizations about the shape, the uniformity, and other things. We have considered the idealized case above that ignored the change in bone properties with temperature change, such as the specific heat of the bone and its thermal conductivity. Somehow Nature has figured out all these things without special computer modeling!

# 14. Cooking Chinese Style

There are at least two good reasons for cutting up meats into small volumes: (1) marinades and spices penetrate more thoroughly into the meat in a shorter time because the inner-volume elements are closer to the surface; (2) smaller chunks cook faster and therefore require less fuel. The faster cooking occurs because (a) the inside of the small cube is closer to the heat source than for a thicker piece and (b) the meat is tumbled during stir-frying, exposing different small surfaces to the higher-temperature direction. The temperature sensed by the meat tends to decrease with distance from the thermal energy source, in this case the pan bottom.

The amount of cooking experienced by a small volume of interior meat is proportional to the temperature experienced and the duration of cooking at this temperature. Both of these quantities are changing during the cooking process. In addition, the thermal conductivity and the thermal heat capacity of the meat are changing because the meat material itself is changing. For example, if the outside becomes charred, its thermal conductivity is significantly reduced, so that the transmission of thermal energy decreases compared to its prior rate. Therefore hamburgers, which must be cooked thoroughly inside to kill the bacteria on the surfaces of the ground-up meat, must never be charred on the outside because the inside will tend to remain uncooked or partially cooked, creating a dangerous eating condition.

The physics here is found in any high school physics text, but the application to cooking was devised millennia ago by ancient chefs who desired a particular result with perhaps a minimum of fuel expenditure. Certainly one can go to the other extreme by slow-roasting a whole carcass in a revolving spit or in a hot ember-lined pit in the ground.

## 15. Baked Beans

The hot beans have taken up water and swelled so that the skins are under great tension. Blowing cool fast-moving air across the beans from pursed lips (instead of an open mouth) lowers their surface temperature and reduces the ambient air pressure. The inside is still hot, so the larger pressure difference results in the hot, high-pressure vapor under the skin pushing outward just a bit more. If the pressure gradient is great enough, the skin will rupture. The slight cooling of the skin material increases the skin tension to reduce the time to rupture.

We can simplify the physics to applying Newton's second law perpendicular to the skin surface. Three forces acting on the skin are important: (1) the inward force of the ambient air pressure, (2) the inward force of the tension in the skin, and (3) the outward force produced by the hot, high-pressure gas within. Blowing the air reduces the ambient air pressure enough to create a net force outward, and skin rupture occurs when the skin

tension has exceeded its elastic limit. During the rupture process, bean skin molecules have been separated because the electrical force holding one molecule to the next has been exceeded.

*Miller, J. S. The Kitchen Professor. Sydney: Australian Broadcasting Commission, 1972, pp. 81–82.*

## 16. Ice Water

The ice at the top brings about faster cooling of the water in the pitcher. As some ice melts, this cold water is more dense than the surrounding water and sinks, cooling the water it passes through. The warmer, less dense water at the bottom is buoyed upward into a cooler region. This mixing helps the water cool faster than when the ice is held at the bottom, because the cold water produced by the ice would remain at the bottom. Thermal conductivity throughout the water would eventually cool the water above, but the convection currents work faster. Of course, vigorous stirring of the ice water eliminates any need for the previous discussion!

One could say that the discussion above is incomplete because in our idealization we have ignored the ice thermal interaction with the ambient air. This interaction can be important, especially on hotter days. The ice does its job when thermally interacting with the water, not with the air! The ice held in the water would be a more efficient direct interaction procedure. So when the ambient air temperature is great enough, they could be competitive.

By the way, this cooling process is exactly the same as the sequence of events that occurs when a pond freezes over in winter. However, in that case, the pond water is prevented from cooling further and from freezing through until all the water reaches 4°C first. This delay in freezing throughout saves the lives of pond organisms through the winter if spring comes soon enough. Evaluated in a different way, there would be no life surviving the worst ice ages on Earth if water did not reach its maximum density at about 4°C!

## 17. Peeling Vegetables

When a tomato is held carefully over a flame and rotated, some of the thermal energy gained by the tomato vaporizes the water just under the skin to locally rupture it. A paring knife can remove the skin easily after the tomato has cooled. Often, just pulling on the ruptured skin is enough. Very hot water can be used instead of a flame, but the effects are not as dramatic, and the peeling is a bit more difficult.

The boiling raises the temperature of the beets to cook them and, simultaneously, a small amount of hot water enters them, so they are slightly swollen. Cold water on the outer surface of the hot beet causes the skin to shrink, but the innards remain hot and swollen. So the stretching skin bursts in several places and becomes easier to remove with a paring knife without being so messy.

The procedure is exactly opposite to placing an ice cube in hot water. Now the outside of the cube tries to expand, but the inside is still cold. One can hear the thermal stresses crack the ice cube.

## 18. Igniting a Sugar Cube

Very small particles tend to ignite more easily. The large surface-area-to-volume ratio for a collection of small particles aids ignition, providing a large combustion area for the chemical interaction of their surface molecules with oxygen and also providing a nearby heat source for sustenance. Therefore, rub the far corner of the sugar cube in some cigarette ash or tiny ash particles from burned paper, then light the ashen cube with the burning match. Ignition is now easy. Oxygen molecules react with molecules in the ash to produce thermal energy and product molecules, including water.

Historically, there have been many examples of the spontaneous ignition of dust particles in the air, such as explosions in granaries where grains crops are stored and in mills that grind grain into smaller particles. A small warm spot in the dusty air, perhaps produced by sunlight, by a match, or by friction, can rapidly spread into a full-scale explosion.

On a less violent scale, simply lighting a campfire outdoors begins with kindling, very small sticks and shavings of wood, which have a very large surface-area-to-volume ratio. The slightly larger sticks can be added once the flame sustains itself. Finally, a whole faggot of sticks can be placed in the firepit to generate a lasting fire.

## 19. Water Boiling

The major thermal effect is the raising of the boiling point (i.e., boiling temperature) by the added salt in solution from 100°C (standard conditions at 1 atmosphere) to about 104°C (if the salt is pure NaCl), a significant change that results in a small time lag before boiling begins again if thermal energy input continues.

In contrast, the actual amount of thermal energy needed to raise the temperature of the sprinkled salt itself is minuscule because the specific heat of NaCl is much lower than for water, and the amount of water by weight in the pot is enormous compared to the amount of salt. The actual boiling

process is more complicated than the simple, ideal version we have considered here, involving nucleate boiling, transition boiling, etc., in the real situation. However, the complete analysis produces the same general argument.

One should ask whether different results would occur with sea salt, a mixture of KCl and NaCl and dead organic matter in larger grain sizes than normal table salt. The slow rate of dissolving the larger sea salt grains may delay the reboiling longer than experienced for NaCl.

The relevant physical data are in most chemistry and physics handbooks as well as in some textbooks. The physics pertinent to the idealized case discussed above is part of traditional physics and chemistry courses, but the more detailed complete analysis can be found only in the technical literature.

## 20. Put the Kettle On

Steam

Spout Gap

No, you cannot see water vapor, that is, water molecules in their gaseous state. If you look closely at the orifice of the spout, there is a clear region perhaps up to one inch long. That's where the water vapor is before it condenses into the steam you *can* see. The temperature of the vapor in the clear region is still too high for droplets of steam to form—that is, collisions of water molecules are too violent to allow them to bind together to form droplets.

In the clear region at the orifice of the kettle, the water molecules are moving so rapidly that when they do collide, the van der Waals attractive force—an induced dipole-dipole electromagnetic interaction—at close range cannot keep them together. As the water vapor cools farther away from the end of the spout, these same collisions produce droplets that grow in size.

## 21. The Watched Pot

Put a pot of water on a flame atop a stove. The thermal energy from the flame raises the water temperature. If the pot has no cover, soon the water vapor pressure above the water surface equals the pressure of the ambient atmosphere. The water is now boiling. If we are high in the mountains, the boiling has occurred at a lower temperature than when we are near sea level. So at higher elevations potatoes may not cook as quickly in the open pot, and the lukewarm water will not make good tea or instant coffee. In the mountains we would be wise to use a pot with a lid so that the total pressure acting downward on the water surface can be higher—vapor pressure plus atmosphere—and so that the water boils at a higher temperature than without the lid, hopefully almost at 100°C.

Suppose the cooking at sea level is done in a pot with a lid. Now the action of lifting the lid to "watch the pot" reduces the thermal energy in the air above the liquid surface as some molecules escape. These escaping water molecules are among the most energetic in the vapor, so they can carry away much thermal energy. The pressure above the liquid is now lower than before, the boiling occurs at a lower temperature, and the cooking takes significantly longer. One should replace the lid and let the food cook undisturbed! Hence the expression "A watched pot never boils." This statement actually refers to the extended cooking time and involves some good physics.

## 22. Ice in a Microwave

Yes! The water molecules in the liquid state rotate a bit in the microwaves and transfer energy to the surrounding molecules to make them jiggle randomly. The water molecules in ice are locked into crystals and are unable to rotate. (Note: The actual details of molecular bonding in the ice are more complicated and show that a minuscule amount of rotation is possible, but an insignificant amount to change the ice to water.) Using microwaves, therefore, one can boil water inside an ice block!

Boiling the water inside the ice block is an example of selective energy absorption. Numerous examples of selective absorption occur in the natural world. For example, the green leaves of plants have chlorophyll A and B molecules that selectively absorb bluish and greenish light for photosynthesis. At an even smaller scale, nuclei are very selective in absorbing gamma rays of specific energies. At the macroscale of meters, we know that rooms can absorb and amplify sound energy at selected resonance frequencies. Some materials are even useful for just the opposite behavior, such as window glass, which has no selective absorption in the visible part of the electromagnetic spectrum. You can take your pick, but the game is played by the rules of nature.

The selective absorption by water molecules (and some other molecules) in a microwave environment is a little different from the other examples given above. At the water molecule's resonant frequencies in the microwave region of the electromagnetic spectrum, the applied field changes so rapidly that very little energy is transferred to the nearby molecules. Microwave ovens actually operate at a frequency that is lower than the frequency at which the absorption is greatest. The food needs to be heated throughout, and by lowering the applied frequency a bit, more microwaves penetrate farther inside, past the outer layer.

Kurti N., and H. This-Benckhard. "Chemistry and Physics in the Kitchen." Scientific American 270, no. 4 (1994): 66–71, 120–123.

Walker, J. "The Secret of a Microwave Oven's Rapid Cooking Is Disclosed." Scientific American 256, no. 2 (1987): 134–138.

## 23. The Glycemic Index

The rate of conversion from one type of molecule to another is a chemical process, with the ratio of surface area to volume for particles in the food being converted being an important factor. Smaller spherelike particles have a higher ratio of surface area (SA) to volume than larger ones. Consequently, since the reactions occur on the surface, material consisting of smaller-diameter spheres convert to sucrose faster than the same material consisting of larger spheres. In fact, for the limiting case of a sphere, the ratio SA/Vol. = 3/R, where R is the radius of the sphere. Physical and chemical processes initiated by the environment occur first at the surface of the particle. In addition, for biological systems, larger particles must be reduced to smaller ones before passing through membranes. So a collection of small particles equal in total mass to one large particle will be reduced to acceptable size faster than the large one because the same amount of chemical solution acts upon a much larger total surface area.

The smaller the particles of ingested food, the faster can be the digestion of the molecules in the intestines because the surface-area-to-volume ratio is higher, and the sooner is the uptake into the blood. The higher temperature used for baking the potato makes its particles smaller than for the same potato when boiled at 100°C, so its glycemic index is greater, reflecting its faster uptake into the blood.

Dates contain some maltose, a sugar that is even faster than glucose in its basic conversion to sucrose in the blood, so their glycemic index is above 100.

A popular book *The New Glucose Revolution* by J. Brand-Miller is available in many libraries and has tables of the glycemic index for numerous foods and some of the recent results from nutrition and diabetes research worldwide. Excerpts from the book and many other resources on the glycemic index and its comparison to the insulin index can be found on the Internet.

# 24. Electric Pickle

Even though the electrical energy source is provided through an AC current, the pickle glows predominantly at one end with a yellowish color that is determined by the pickling solution and the pickle type. Reliably predicting which end will glow has not been achieved. There is no actual symmetry here in the shape or chemical composition of the pickle, so alternate glowing is less likely and never seen. The conjecture is that the pickle is now acting like an electrical diode, passing current in one direction only!

The authors listed below performed an experiment by taking a visible light spectrum of the glowing pickle, using a spectrometer with a diode array detector. A fiber-optic probe was used to channel the yellow glow to the spectrograph, and a calibration spectrum was taken of a sodium chloride flame test. The emission spectra of the two are nearly identical

This pair of emission lines, at 589.00 nanometers (nm) and 589.59

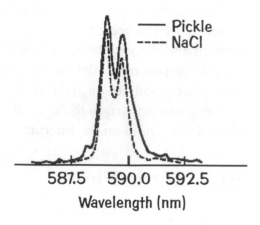

- Pickle
- - - - NaCl

587.5    590.0    592.5
Wavelength (nm)

nanometers, indicate a characteristic of sodium emission, called the sodium D line doublet. Josef Fraunhofer observed these lines in the emission spectrum of the Sun, in about 1817. We know now that these lines are due specifically to an electronic transition of sodium atoms in the gas phase.

The pickle conducts electricity due to the vinegar (acetic acid) and sodium chloride salt used to make it. Sodium ions in the pickle liquid attach electrons from the flowing current. These ions are neutralized electrically, forming excited sodium atoms in two different excited electronic states (hence the emission doublet). Because of the heat and sparks and general pandemonium around the electrodes stuck in the pickle, these sodium atoms are in the gas phase. They emit yellow light as they relax to the ground state.

Appling, J. R., F. J. Yonke, R. A. Edgington, and S. L. Jacobs. "Sodium D Line Emission from Pickles." The Journal of Chemical Education 70, no. 3 (1993): 250.

# 25. Space-Age Cooking

Unlike electric cooktops, which generate thermal energy by the electrical resistance of the burner coils, magnetic induction cooktops generate thermal energy by the magnetic resistance of the metal cooking vessel itself. The 60 Hz AC current flowing in the induction coil beneath the ceramic surface produces an alternating magnetic field than interacts with the Fe atoms—for example, in the iron frying pan—to oscillate its magnetization 120 times per second. The magnetization direction changes have resistance, so much energy goes into thermal energy in the metal of the pan. Iron and stainless steel pans will work, but aluminum, copper, glass, and ceramic pans and pots will not. The advantages are no noise and no hot cooktop except where the pan has been in contact.

Cooking with light is not done by lasers! *Light* is meant in the broader sense of the word, the infrared (IR) through the visible into the ultraviolet (UV) part of the electromagnetic spectrum. Banks of 1500-watt halogen lamps in the oven walls put out about 70 percent IR, 10 percent is visible light, and the remaining 20 percent is simply heat. The IR is not thermal energy, but when IR is absorbed by molecules, their random motions can be increased. Thermal energy ("heat" in the vernacular) is the random kinetic energy of molecules and atoms. These frequencies penetrate meat only about half an inch at most. Thermal conduction transfers some of this thermal energy farther inside.

However, these light ovens also have a microwave source to penetrate with microwaves to cook the interior of the meat. So while the outside is being browned by the light, the inside is cooking via microwaves. The overall benefit is much faster cooking than is possible in conventional ovens.

*Wolke, R. L.* What Einstein Told His Cook: Kitchen Science Explained. *New York: W. W. Norton, 2002, pp. 303–307.*

# Chapter 2
# Does Anybody Really Know What Time It Is?

## 26. January Summer

Yes, the Northern Hemisphere enjoys summer in January quite often (in the cosmic scheme of things), repeating, every 25,800 years, the period of Earth's precession. Just like a top with its axis precessing, Earth experiences a precession of its axis with respect to the stars with a 25,800-year period of oscillation. So every 12,900 years the

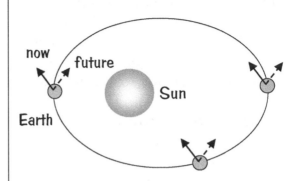

North Pole will be alternating from the extreme of being pointed toward the Sun and to being pointed away from the Sun in January. At present and for some years to come, the North Pole points away from the Sun when Earth is at the perihelion position in its orbit on about January 5 each year. Gradually over the next 12,900 years the North Pole will precess around to receive more and more radiant energy in January.

However, we need not wait nearly so long because the ellipse of Earth's orbit is also precessing, so our summer will coincide with perihelion in only about 10,000 years!

## 27. Proximity of Winter Solstice and Perihelion

The proximity of the two dates is an artifact of the particular century we live in. The date of perihelion does not remain fixed, but slowly moves later into the year at the rate of about one full day every 58 years. It turns out that the period from perihelion to perihelion (the anomalistic year) is about 25 minutes longer than the year defined from equinox to equinox (the mean tropical year). The date of perihelion thus moves completely through the tropical year in about 21,000 years. This slow change in the date of perihelion may have a long-term effect on Earth's climate. At this time the temperature extremes are moderated somewhat in the Northern Hemisphere, but that will change as the perihelion shifts in the direction of summer.

# 28. Earth's Speed

Since perihelion occurs in early January, Kepler's second law implies that Earth is traveling faster during the winter months. The time for Earth to travel from the autumnal to the vernal equinox, taken as a fraction of the year (T = 178.83/365.25), can be used to find an accurate value of the eccentricity of Earth's orbit, $\varepsilon = 0.5\,\pi$ (0.5 − T) = 0.01632, about 2 percent away from the precise value of $\varepsilon$ = 0.016713. A more accurate formula based on T is found in the reference below.

*Snyder, R. "Kepler's Laws and Earth's Eccentricity."* American Journal of Physics 57 (1989): 663–664.

# 29. The Equinox Displaced

On the dates of the equinoxes, the day is about seven minutes longer than the night at latitudes up to about 25 degrees, increasing to 10 minutes or more at latitude 50 degrees.

The moment of the equinox occurs when the geometric center of the Sun's disk crosses the celestial Equator. But the standard definition of sunrise is the time when the Sun's upper limb is just breaking the horizon, and sunset when the Sun's upper limb is just disappearing below the horizon. This adds one Sun semidiameter (about 16 arc min.) at both sunrise and sunset, extending the duration of daylight by a little over two minutes.

The other factor is atmospheric refraction, which causes the rays to bend around the horizon. As a result, we see the Sun about 34 arc minutes higher at both sunrise and sunset, adding roughly four minutes to the time that the Sun is above the horizon.

In spring the days get longer as we approach March 20, and the date of equal day and night occurs several days before the March equinox, about March 17 at latitude 40 degrees.

Conversely, in the fall it takes several extra days for the time when the Sun is seen above the horizon to shrink to 12 hours. The date falls on about September 26 at latitude 40 degrees.

On their website, the U.S. Naval Observatory publishes excellent sunrise and sunset tables for any location.

# 30. The Dark Days of December

There are two effects that, together, determine the local times of sunrise and sunset. One is called the equation of time; the other is the Sun's declination.

Earth's orbit around the Sun is slightly elliptical. As a result, the speed of the Sun's apparent motion across the sky is a bit faster in winter than in summer. Clocks, however, run at a constant speed, so there is usually a discrepancy—up to 16 minutes—between clock time and the solar time shown by a sundial. We refer to this discrepancy as the equation of time.

The Sun's declination, its angular distance above or below the celestial equator, determines the maximum height of the Sun in the sky on any given day, thus causing our seasons. In late December, the daily rate of change of the Sun's declination is rather small. It is, in fact, exactly zero at the December solstice ("solstice" means "sun stationary"). Hence in late December, or more precisely from about December 8 to January 5 at latitude 40 degrees

north, the equation of time has the dominant influence over the changes in sunrise and sunset times. Prior to December 8, however, the declination effect is dominant, pulling the sunset to its earliest time on December 8. Then the equation of time takes over, and during the two weeks before winter solstice all the shortening of the day comes from the later clock time of sunrise. After winter solstice the days lengthen, even as the sunrises continue to get later until January 5.

Steel, D. Marking Time: The Epic Quest to Invent the Perfect Calendar. *New York: John Wiley & Sons, 2000.*

# 31. Days of the Year

While the time interval to return to its same point in the orbit is 365.2422 days, Earth executes 366.2422 rotations on its axis. One can demonstrate this result by taking two coins, holding one in place on a table, and rolling the second coin in contact with the fixed coin without slipping. Therefore the number of solar days is 365.2422, but the number of sidereal days (i.e., with respect to the stars) is one more for one orbit of the Sun.

# 32. Leap Years

In years divisible by four, every four years is a leap year except years divisible by 100. If the mean interval between vernal equinoxes, called the tropical year, lasts 365.2422 days, then in 100 years we should experience 36,524.22 days. But there will be 24 leap years in a century normally, so there will be 0.22 day left over. So every 400 years is declared to be a leap year with one extra day to approximate the 0.88 day. The year 2000 was the first such leap year on a year divisible by 100 since the modern calendar began general use in the late 1600s. By the time the British were ready to go along with the rest of Europe in the 1700s, the old Julian calendar required a correction of eleven days! The Gregorian calendar was adopted in Britain in 1752, with Wednesday, September 2, 1752, being followed immediately by Thursday, September 14, 1752.

The famous physicist Isaac Newton was born on Christmas Day, 1642, on the Julian calendar but on January 4, 1643, on the Gregorian calendar in use today. Therefore, Newton was not born in the year of Galileo's death, 1642!

# 33. Full Moons

No, the orbital period of the Moon is 27.554 sidereal days, and the average interval between full moons was 29.535 days for the twentieth century. The difference between these two time periods occurs because Earth is moving with respect to the stars, so the Moon must travel slightly farther around its Earth orbit to reach its full Moon position along the Sun-to-Earth radial line.

*U.S. Naval Observatory, Nautical Almanac Office.* The Astronomical Almanac for the Year 2000. *Washington, D.C.: U.S. Government Printing Office, 2000.*

# 34. Moon Time

The person is at the desk at 12:20 during the noon hour, not at night. The Sun must be at the upper left because the Moon is illuminated from this direction. This daytime Moon is seldom noticed because the sky is normally bright, but the daytime Moon is up as often and as long as the Moon at night.

Perhaps the easiest way to appreciate the appearances of the Moon at night and during daylight is to use a lamp for the Sun and two spheres, one representing Earth and the other the Moon. Fix the lamp position and the Earth position, but move the Moon around Earth to observe its illumination phases. Stop the motion at several points in the orbit of the Moon to observe its view from daytime locations on Earth. You also might reconstruct the scene in the diagram.

*Pryor, M. J. "Phases, Models, and Cartoons."* The Physics Teacher 3, *no. 6 (1965): 264, 288.*

# 35. Lunar Calendar

Modern farming methods tend to plant crops at approximately the same time year after year, with minor adjustments for quirks in the weather. Hence, a particular crop is usually planted about 365 days after its planting the previous year, plus or minus about 10 days. One example is spring wheat, usually planted on about April 15 in the northern Plains states of the United States.

Rice is a different kind of organism than wheat as far as its environmental needs. Rice planted at about the same date every year will sometimes permit two good rice crops per year, but in most years the farmer will get only one good rice crop. The cause is the sometimes detrimental appearance at night of the full Moon, which can interfere with the growth cycles of the rice plant.

By planting rice according to the same date on the lunar calendar instead of the solar calendar, farmers can often harvest two good rice crops every year. The young rice shoots are very sensitive to the light intensity at night during their photoperiod-sensitive stage, so the timing of the Moon's brightness is essential for a good crop. Because the lunar calendar shifts with respect to the solar calendar dates each year, the solar calendar provides bad timing for planting rice.

The photoperiod-sensitive stage occurs before panicle initiation (where the seed parts develop) and may vary extremely from one variety to the next, from days to months. Photoperiod sensitivity is a natural mechanism based on the plant's ability to distinguish precise differences in the ratio of day length to night length. The biological mechanism causing photoperiod sensitivity is quite complex and involves several genes. Essentially, some varieties should be planted only during certain times of year to ensure that prevailing day length/night length conditions will trigger panicle initiation when desired.

Many varieties of rice are sensitive to bright moonlight, which can interrupt their growth sequence. However, newer varieties have been bred and others will be genetically modified to decrease their light sensitivity during critical photosensitivity times so that the moonlight will have a minimal effect.

The use of the Moon for timing of events is not restricted to rice farming and its related festivals worldwide. For Christians, Easter Sunday is the first Sunday after the first full Moon after the vernal equinox!

*University of the Philippines College of Agriculture in cooperation with the International Rice Research Institute (IRRI), comps.* Rice Production Manual, *rev. ed. Manila: compilers, 1970.*

*Yano, M., et al. "Hd1, A Major Photoperiod Sensitivity Quantitative Trait Locus in Rice, Is Closely Related to the Arabidopsis Flowering Time Gene Constans."* Plant Cell 12 *(2000): 2473–2484.*

# 36. The Sandglass

The hourglass shape ensures that the time scale on the glass is uniform, with equal distances between scale divisions corresponding to equal time intervals. If the sandglass didn't taper, the top of the sand column would descend at increasing speed. We can mathematically determine the proper shape. The speed, $V$, at which the sand is escaping from the opening is given approximately by Torricelli's formula $V \sim \sqrt{(2gy)}$, where $g$ is the local acceleration of gravity and $y$ is the height of the sand column in the upper glass. Let $A = \pi r^2$ be the circular cross-sectional area of the upper glass at the top of the sand column and $v$ the speed at which this top is falling, then $Av = aV$, where $a$ is the area of the opening, because the sand is approximately incompressible. Substitution produces $y \sim cr^4$, with constant $c = \pi^2 v^2/(2ga^2)$. A plot of $y$ versus $r$ will produce the familiar hourglass shape.

Sandglasses *without* time interval markings have been used since before the fourteenth century to time speeches at town meetings and other events. When the sand had run its course, the speaker's time was done. They are still used today in some board games and as kitchen timers. Sandglasses with ruled markings haven't been so popular, being replaced early on by mechanical watches and clocks.

Jargocki, C. P. Science Braintwisters, Paradoxes, and Fallacies. *New York: Charles Scribner's Sons, 1976, pp. 6, 70.*

# 37. Old Watch

The old watch will run fast. The balance wheel is the basic component that oscillates "exactly" 300 times for each minute on the watch face—that is, the wheel changes direction 10 times each second! The moment of inertia of the balance wheel depends on how much ambient air is dragged along during each oscillation. The source of energy is a wound spring that is essentially unaffected by the air because of its extremely small change in configuration.

In the mountains, the viscosity and density of the air decrease slightly, allowing the balance wheel to oscillate faster. Newton's second law applied to this rotational motion is required. From its momentary stop to change direction, the balance wheel must accelerate to its maximum angular velocity, then accelerate back to rest, etc. The net torque $\tau$ equals the moment of inertia $I$ times the angular acceleration $\alpha$—that is, $\tau = I\alpha$. The moment of inertia is determined by the mass distribution with respect to the rotation axis, and the air carried along with the balance wheel motion adds to the moment of inertia of the wheel alone. That is why the watch must be recalibrated when the location of the owner is at a different elevation than the factory.

When the balance wheel drags along less air mass at a higher elevation, the moment of inertia is less for the same net torque, so the angular acceleration is greater. Less time is needed to reach top angular speed, and less time is needed to come to rest again.

*Jargocki, C. P.* Science Braintwisters, Paradoxes, and Fallacies. *New York: Charles Scribner's Sons, 1976, pp. 6, 71.*

# 38. Reading a Digital Timer

For a digital timer that displays the elapsed time to one-hundredth of a second, the minimum uncertainty in the interval depends on the software and/or hardware method used to display the last digit. Suppose that the hundredths digit fraction from 0.00 through 0.49 is displayed as zero hundredths and from 0.50 through 0.99 is displayed as one hundredth. Likewise, 1.00 through 1.49 is displayed as one hundredth, and 1.50 through 1.99 as two hundredths. Seeing a 1 in the hundredths place then corresponds to the range from 0.50 through 1.49, so the minimum uncertainty in the elapsed time value is ± 0.50, or one-half of one-hundredth of a second. The reported elapsed time should be given

as, say, 3.45 seconds ± 0.005 second, which is an awkward notation, because the display goes only to hundredths of a second, yet the uncertainty is smaller. Therefore, by agreement, the elapsed time is given as 3.45 seconds ± 0.01 second so that the number of decimal places is the same for the value and the uncertainty— that is, there is an uncertainty of plus or minus one digit in the smallest time interval position of the display.

# 39. Eternal Clocks?

The laser clocks and the atomic clocks must maintain a vacuum within a reasonably small range of parameters to function accurately. The temperature and pressure must be maintained within a certain tolerance because temperature fluctuations or pressure fluctuations could bring about inaccuracies. Even the outgassing of atoms and molecules from the container walls can create severe problems for some designs. Certainly, improvements will be made to ensure longer lifetimes and more robust timepieces. But maintaining vacuums, low temperatures, and so on, for decades and centuries will be constant problems in these sophisticated systems.

Whether a 10,000-year (or even a 1,000-year!) mechanical clock will ever exist and can stand the test of time and environment is doubtful. A group of engineers and futurists are presently developing such a clock containing a stack of rotating metal rings connected to a torsion pendulum. Periodic winding will be required, perhaps once a year or so.

No special environment is needed, although the assumption seems to be that a standard atmosphere with limited pollutant content is a reasonable expectation. We do know from a variety of scientific research projects that the oxygen concentration in the atmosphere has been very nearly constant at 21 percent for millions of years, so the rusting rate of the exposed metal can be predicted. But we do not know what future chemistry will bring. Even a local environmental disaster such as excess acidity in the air from volcanic eruptions, a chemical explosion, or careless disposal could shorten the life span of the clock dramatically.

*Gibbs, W. W. "Ultimate Clocks."* Scientific American *287, no. 3 (2002): 86–93.*

# 40. Room Light

For a nanosecond flash, the light pulse length $d = 30$ cm is calculated with $d = ct$, where c is the speed of light and $t$ is the time interval. When summing all the entering light, the photodetector displays an initial rise from the light scattering from nearby walls with increasing intensity until maximum, and then a decrease to zero after the light from the far-corner reflections is received. The detailed intensity curve could be simulated on a computer.

With the photodetector array, the image shows the six nearest spots—the centers of each equidistant flat surface—which grow bigger and then form rings of reflected light, then many arcs of light until eight corners appear and disappear.

When the flash length is extended to 1 microsecond, the light pulse is 300 meters long. There will be an initial detector response rise and the rings of light from the walls will be seen for a very small fraction of the total imaging, then flooded, then decreased.

We are not accustomed to light pulses lasting for milliseconds or less in daily life. But even nanosecond pulses are very long in some research fields. For example, femtosecond ($10^{-15}$ second) and shorter light pulses are used in chemistry to watch molecular interactions in progress. The present record for subdividing the second with a laser strobe light is a few hundred attoseconds, an attosecond being a billionth of a billionth of a second!

# 41. Right to Left Driving Switch

Yes, as long as the accelerations and decelerations required are within the normal driving ranges, there should be no problems. One could check out each roadway with physical trials, or one could make an aerial video and play back the video in the reverse direction, essentially reversing time. If the car accelerations appear unusual in the reverse sequence, there will be driving problems.

Even nature at the most fundamental level is cognizant of left versus right, a surprising discovery in the 1950s related to the weak interaction, one of the four fundamental interactions, the other three being the gravitational, the electromagnetic, and color (also called the strong interaction). For the latter three, the interaction strengths are the same for left-hand spinning particles and right-hand spinning particles. But for the weak interaction, the evidence shows that nature actually excludes any weak interaction behavior for a right-hand spinning particle! The origin of this behavior biased toward left-hand spinning particles is described by the mathematics in the standard model of leptons and quarks.

*Bartlett, A. A. "A Simple Problem from the Real World That Can Be Solved through Time Reversal."* American Journal of Physics 42 (1974): 416–417.

## 42. Light Clock

Yes and no. The clock will continue to keep accurate time in both reference frames, in the laboratory frame, and in the rest frame of the clock. But the tick rates will be different. In the frame moving with the clock, the light flashes follow the same path as before, reflecting perpendicularly to each mirror, keeping the same tick rate.

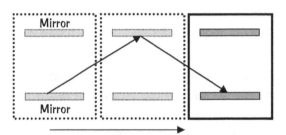

In the laboratory frame, the light continues to reflect off each mirror repeatedly, but during transit from one mirror to the other the path length is longer, being the diagonal of a right triangle. If the speed of light is the same value in both reference frames, then the time interval between reflections (and clock ticks) will be longer in the lab frame now than for the clock at rest. Therefore, a moving clock ticks slower than an identical clock at rest. And this phenomenon is true for all clocks, no matter how they are made.

A more complicated situation occurs when the light clock is *accelerated* parallel to the mirrors. We suggest that you think about this case when there is ample time in your schedule.

*Feynman, R. P., R. B. Leighton, and M. Sands.* The Feynman Lectures on Physics. *Vol. 1. Reading, Mass.: Addison-Wesley, 1963, pp. 15-5–15-7.*

# 43. Time Reversal

Answer b: the acceleration is still downward. The reversed motion is upward, but the object is decreasing its speed because the acceleration is downward. A good example of this behavior is the flight of a ball tossed upward. At all moments the acceleration is downward, toward Earth's center. Yet the ball moves upward with decreasing speed, turns around, and moves downward. Even at the turnaround point its acceleration is downward.

Quite often people become confused between velocity and acceleration. They are two different vector quantities that should be separated conceptually, but they are mathematically related. Their directions can be the same or opposite along the line of motion. Newton's second law relates forces and accelerations but says nothing about velocities, for example. And we know that Aristotle was wrong when he proposed that a force was required to keep an object moving. The real world operates with just the opposite rule because *no* net force is required to keep an object moving in a straight line at a constant speed!

# 44. Molecular Clock

The changes in the DNA during the evolution of organisms do not occur at a common rate because any change in a critical essential protein coding does not produce a viable organism even though a change in the nonessential DNA might—that is, changes in the DNA that do not affect the biochemistry critically will be tolerated. There are vast regions of DNA where such ineffective changes can occur, but any change in the other regions programmed for the production of essential biomolecules will be disastrous.

If we assume the ideal case, that in principle the changes would be equally probable at any random location along any DNA chain, and we assume that the organism will grow and reproduce the next generation, then we could have a molecular clock. However, as we know, just as all genes are not equal in value at any given time, all DNA sequences are not equal in value. In particular, some DNA sequences code for proteins that control the expression of other DNA genes themselves, turning them on and off at appropriate times in the development of cells in the organism. The Hox protein in insects, for example, will determine the structure of several different body parts, and slight changes in its amino acid sequence are major contributors to insect evolution. Therefore, both complementary DNA strands at the critical location need not be affected for the appearance of obvious phenotype changes.

However, the whole DNA mechanism and its subsequent biochemistry in the cell are much more robust than originally realized. The fact that many of the amino acids have several DNA base code triplets of nucleic acids for their selection is a built-in resiliency that can produce a viable organism even when the DNA has an error of this kind. In addition, if the erroneous amino acid substitutes at a location that is not critical for the 3-D shape and the operation of the protein, once again there is a built-in resiliency. Let's face the fact that Nature is much more clever than we can ever hope to be!

Gibbs, W. W. "The Unseen Genome: Gems among the Junk." Scientific American 289, no. 5 (2003): 46–53.

Ronshaugen, M., N. McGinnis, and W. McGinnis. "Hox Protein Mutation and Macroevolution of the Insect Body Plan." Nature 415 (2002): 914–917.

# 45. SAD

Yes, if they suffer from SAD. At first one might think that the variation in the length of day and night changes so little from January to June that no one living at the Equator would suffer from SAD. This reasoning is true if everyone went to bed at sunset and arose at sunrise. The increasing light at sunrise would trigger the start of another circadian rhythm that brings about biochemical changes in our bodies.

But even people living near the Equator are tuned no longer into the rise and setting of the Sun. There are problems in their circadian rhythms because the bright artificial lighting means that people stay awake long past sunset, delaying and shifting the maximum in certain biochemical cycles beyond their evolutionary time of day. In particular, greenish light from televisions and clock radios passes into the eyes, even through closed eyelids while asleep, to trigger the pineal gland to initiate some of the biochemical circadian rhythm shifts.

Wright, K. "Times of Our Lives." Scientific American 287, no. 3 (2002): 59–66.

# 46. Two Metronomes

For the case of the periodic perturbation of one metronome by the other, the mode-locking occurs when the perturbing frequency is sufficiently close to the unperturbed frequency of the metronome. When a metronome is placed on the skateboard, the movement of the pendulum causes the skateboard itself to move slightly, usually in the opposite direction to the pendulum swing, since the metronome base is kept in place on the skateboard by static friction or could be bolted down. Some of the energy of the metronome base motion is transferred to the skateboard, and this very small amount of energy is further transferred along the skateboard in several directions, with some amount reaching the other identical metronome.

If at first this energy arrives at some random phase point in the oscillation of the second metronome, eventually its regular energy delivery becomes more and more effective in synchronizing the pendulum oscillations. Of course, the second metronome is acting on the first metronome in the same way simultaneously. The synchronization is normally in-phase, but antiphase synchronization can occur in special conditions. (See the second reference below for details.)

The behavior can be represented by two equations for two harmonically driven oscillators with a significant amount of dampening. If the dampening were not significant, then we would see two coupled pendulums alternating their swing behavior out of phase from maximum amplitude to nearly zero amplitude. In the actual case, the pendulums simply synchronize and keep nearly identical time.

For the case of one of the pendulums being driven by a force random in time, their fluctuating behavior can converge to an identical response. Both pendulums would exhibit the same random fluctuations eventually. For both periodic and aperiodic driving forces, asymptotic stability results for linear oscillators properly damped. That is, small changes in the parameters of the linear oscillator or the driving force result in only small changes in the asymptotic behavior. The equation of motion for each oscillator is mathematically equivalent to describing a linear spring in a viscous medium with a fluctuating driving force.

According to the first reference below, the mode-locking can occur also for a wide range of aperiodically driven nonlinear oscillators in the physical and biological sciences, from nonlinear electrical circuits to neural systems. As in the periodically driven systems, the synchronization of randomly driven nonlinear oscillators was found to be structurally stable, which means that even in the presence of small amounts of noise an approximate synchronization is achieved.

Jensen, R. V. *"Synchronization of Driven Nonlinear Oscillators."* American Journal of Physics 70 (2002): 607.

Pantaleone, J. *"Synchronization of Metronomes."* American Journal of Physics 70 (2002): 992–1000.

# 47. Time Symmetry

No, nature does not need to obey time symmetry at the most fundamental level. Equations sometimes have more symmetry than the actual underlying physics behavior. For example, even though the tensor equations of general relativity are time-symmetric, they can be derived from a more fundamental type of mathematical entity called a twistor. Twistor equations of general relativity are not time-symmetric. In applications to a black hole, for example, the tensor equations predict time symmetry, but the twistor ones do not. As a consequence, the formation of a black hole and the time-reversed version cannot both represent real physical behavior.

One might think that the quantum theory described by the Schrödinger equation is time-asymmetric, the equation being first order in time. As Roger Penrose points out in the reference below, quantum theory and its equations are indeed time-asymmetric. The wave function can be used to calculate the probability of a *future* state on the basis of a known *past* state, but not the other way—that is, one cannot calculate the probability of a past state on the basis of a future state. You cannot retrodict the past!

Hilgevoord, J. "Time in Quantum Mechanics." American Journal of Physics 70 (2002): 301–306.

Penrose, R. The Emperor's New Mind. Oxford: Oxford University Press, 1989, pp. 354–359.

# Chapter 3
# Crazy Circles

## 48. Spider and Fly

To find the shortest path between any two points on a cube not on the same face, one convenient method is to lay out the six faces of the cube on a plane

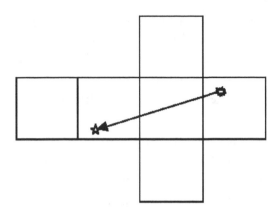

and draw the straight-line path. Of course, all parts of the path must be on the faces, and the appropriate faces sharing a common edge must retain their relative positions and orientations.

Steinhaus, H. Mathematical Snapshots, 3rd ed. New York: Oxford University Press, 1983, pp. 173–176.

## 49. Moon Distance

The laser light pulse traveling from the Earth to the Moon and back will encounter the Earth's atmosphere twice. The pulse will have an initial known rise time and decay time, but these times will be extended by passage through the air medium. We assume ideal reflection at the Moon's corner reflector—that is, no pulse spreading in angle or in time.

First, the ideal case. We assume that the laser source and the reflector on the Moon are opposite each other on the line connecting the centers of Earth and Moon and that Earth's atmosphere does not affect the transit time. The major source of uncertainty will be the ability of the detection system to locate the half-height point on the rise time of the outgoing pulse and the same point on the incoming pulse. If the system is good to about a picosecond in detecting this point, then a transit time of 2.56 seconds for the $3.84 \times 10^8$ meter distance corresponds to a timing uncertainty of better than one part in 100 billion, with an uncertainty in distance of less than 4 millimeters. That is, with the proposed laser system, one can measure the distance to the Moon to almost the same distance uncertainty as one can measure the length of a table with a meterstick!

Of course, the atmosphere will foul things up a bit. The index of refraction and the change of this index with altitude will both alter the light pulse speed and spread out the pulse rise time and rise shape. Sophisticated signal processing techniques can eliminate most of these atmospheric effects. So the final uncertainty will be determined in the electronics creating the laser pulse and detecting the arrival of the pulse's leading edge.

# 50. Ideal Billiards Table

On this ideal billiards table for which the incident and reflection angles at the cushion are equal, one simply considers adjacent mirror image copies of the table in a gridwork. Imagine placing the ball in appropriate adjacent tables at mirror image positions; then draw the straight line to the pocket to find the collision points on the cushions.

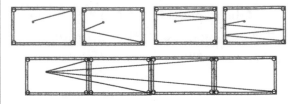

Normal billiards and pool tables are marked around the perimeter to accomplish precision bank shots. Using these markers takes practice. These real tables with their markings are not very similar to the ideal table discussed above. Moreover, the rolling ball before the collision with the cushion may have additional spin—"English"—about an axis not parallel to the table and perpendicular to the travel direction. The professional player uses all these quantities in the particular shot, but we amateurs simply enjoy the results as we practice more of the many possible improvements to our games.

Steinhaus, H. Mathematical Snapshots, 3rd ed. New York: Oxford University Press, 1983, pp. 61–64.

# 51. Wallpaper Geometry

Standing inside the cube, to your right you see your left side of your person in the cube to the right. To your front you see your back. To the top you see the soles of your shoes. You see a 3-D array of yourself from many different views, at many different distances, at many apparent sizes, and at many different image intensities. This view is not like being inside a cube with reflecting mirrors on all sides because no image is reversed.

Cosmologists are trying to determine whether our 3-D space is mathematically and physically discrete—that is, compartmentalized into large cubes or regular dodecahedrons, each being perhaps as large as 10 billion light-years or bigger. If so, seeing a galaxy in one direction could be complemented by seeing the same galaxy in the opposite direction. Of course, several problems exist, such as the distance being greater in one direction than in the other, with the consequence that the galaxy is being seen not only from the other side but also at a different time in its evolution of structure. There may be multiple copies of the galaxies to confound things. There might even be multiple copies of each of us! Any positive results will bring about a revolution in our thinking about space and time in the universe.

Levin, J. How the Universe Got Its Spots: Diary of a Finite Time in a Finite Space. Princeton, N.J.: Princeton University Press, 2003, pp. 132–155.

Thurston, W. P., and J. R. Weeks. "The Mathematics of Three-Dimensional Manifolds." Scientific American 251, no. 1 (1984):108–120.

# 52. Space-Filling Geometry

First consider a two-dimensional flat space. A plane tesselation (or two-dimensional honeycomb) is an infinite set of polygons fitting together to cover the whole plane once, with every side of each polygon belonging to just one other polygon. A regular tesselation has regular polygons. There are three regular tesselations of the plane: equilateral triangles, squares, and regular hexagons. There are additional plane tesselations with two or more convex polygon shapes. One also can cover the plane with Penrose tiles, polygon pairs with at least one polygon not being convex.

Now consider an additional spatial dimension. A three-dimensional honeycomb (or solid tesselation) is an infinite set of polyhedrons fitting together to fill all space once, so that every face of each polyhedron belongs to one other polyhedron. If we require all the polyhedrons to be identical, then the only regular honeycomb is the one filled with cubes, eight at each vertex. If we allow two different regular polyhedrons, one can fill the space with eight regular tetrahedrons and six regular octahedrons surrounding each vertex. These space fillings and others determine many of the natural crystal structures.

From the apparent simplicity of a 3-D space filled with cubes, one may think that this solid tesselation would be the most likely mathematically if real space is discrete instead of continuous. However, mathematicians can show that the most likely and interesting 3-D discrete space is the non-Euclidean tesselation by dodecahedrons, of which there are two kinds, depending on the angle of twist in relating one dodecahedron to the adjacent one. For further information see the Thurston and Weeks reference below.

Coxeter, H. S. M. Regular Polytopes. *New York: Dover, 1973, pp. 58–74.*

Levin, J. How the Universe Got Its Spots: Diary of a Finite Time in a Finite Space. *Princeton, N.J.: Princeton University Press, 2003, pp. 132–155.*

Thurston, W. P. and J. R. Weeks. "The Mathematics of Three-Dimensional Manifolds." *Scientific American 251, no. 1 (1984): 108–120.*

# 53. Archimedes' Gravestone

Archimedes (287?–212 B.C.E.), perhaps the greatest mathematician of ancient times, was the first to calculate the volume ratio of the sphere inside the cylinder. With a sphere and a cone inside the cylinder touching top, bottom, and sides, Archimedes determined that their volumes are in the ratios 1:2:3! 

The Roman general, Marcellus, tells of how he searched for and found Archimedes' gravesite with this headstone. Archimedes was killed in 212 B.C.E. during the capture of Syracuse by the Romans in the Second Punic War after all his efforts to keep the Romans at bay with his machines of war had failed. Plutarch recounts three versions of the story of his killing that had come down to him:

1. "Archimedes was, as fate would have it, intent upon working out some problem by a diagram, and having fixed his mind alike and his eyes upon the subject of his speculation, he never noticed the incursion of the Romans, nor that the city was taken. In this transport of study and contemplation, a soldier, unexpectedly coming up to him, commanded him to follow to Marcellus; which he declining to do before he had worked out his problem to a demonstration, the soldier, enraged, drew his sword and ran him through."

2. "A Roman soldier, running upon him with a drawn sword, offered to kill him; and that Archimedes, looking back, earnestly besought him to hold his hand a little while, that he might not leave what he was then at work upon inconclusive and imperfect; but the soldier, nothing moved by his entreaty, instantly killed him."

3. "As Archimedes was carrying to Marcellus mathematical instruments, dials, spheres, and angles, by

which the magnitude of the Sun might be measured to the sight, some soldiers seeing him, and thinking that he carried gold in a vessel, slew him."

Archimedes was buried in Syracuse, where he was born, where he grew up, where he worked, and where he died. On his grave there is an inscription of $\pi$, his most famous discovery. Also placed on his tombstone is the figure of a sphere inscribed inside a cylinder and the 2:3 ratio of the volumes between them, the solution to the problem he considered his greatest achievement.

His nicknames were, "the Wise One," "the Master," and "the Great Geometer."

*Plutarch.* Lives of Noble Grecians and Romans. *Translated by A. H. Clough. New York: Random House, Modern Library, 1992, p. 517.*

# 54. Brain Connections

A million neuron model of the brain is still quite a formidable programming task for a computer simulation, but there would be no information transfers from neuron to neuron. Why not? Because any neuron in this model of the human brain would have on average nearly zero inputs. One calculates as follows: if $10^{11}$ neurons have 1,000 connections each, say, then the average is 1 connection per $10^8$ neurons.

Therefore, a model with $10^6$ neurons will not work as a useful scale model of the real brain.

Of course, one could simply take a small volume of the brain containing 1 million neurons and ignore connections to other parts. Or one could artificially modify the unusable computer model above by ensuring a few connections or more to each neuron. Whether the behavior that ensues is realistic must be determined. The more practical approach is to model a small section of the brain—perhaps tens of thousands of neurons and all their interconnections—in a focused study and simulation. A grid of computers, each representing one small section, could then be used to simulate a larger portion of the brain. Hopefully, when quantum computers become a reality, they will be able to simulate the whole brain. Whether the brain behaves quantum mechanically and requires quantum superposition for its operations is presently unknown.

There is the remarkable problem of information storage in the brain—that is, where exactly is information stored? If each neuron stores only one bit of information, then the human brain is not large enough by many factors of ten! In 1989 Roger Penrose suggested that each neuron must be capable of storing many bits of information, in contrast to the prevailing ideas. Subsequently, the numerous microtubules in each neuron were found to participate in the information storage game. There still remains the question of what each stored bit of information represents.

Penrose, R. Shadows of the Mind. *Oxford: Oxford University Press, 1994, pp. 358–377.*

# 55. Configuration Space

There are many ways to approach this problem of describing the arm position in physical 3-D space. We consider one approach only. In all approaches, the end of the rodlike hand must touch the specified point, so three numbers define the end point of the hand.

Let's start at the fixed shoulder position. Two numbers will describe the upper arm position, the angle in the vertical plane measured from a fixed vertical axis through the shoulder, and an angle about this vertical axis. Two more numbers describe the forearm position, an angle in the vertical plane measured from a vertical axis through the end of the upper arm, and an angle about this vertical axis. Likewise, two more angles are needed for the hand.

At least six numbers are necessary for the robot to locate the particular point in the room. The program will calculate the extent of the arm to determine its end point distance, thus requiring three more numbers, the lengths of the three parts. The space of operation is nine-dimensional and is called a 9-D configuration space to distinguish between physical space and coordinate space. Of course, one could have determined this result by realizing that each rod end point requires three coordinate values to be specified.

The movement of the arm to touch the point is the next challenge. If feedback exists in the robot, such as visual feedback of the hand position and the desired point location, the movement algorithm can use a correction procedure that becomes finer and finer as the fingertip approaches the point, as we humans tend to operate. If there is no continual feedback mechanism, then the algorithm must move the arm to the point directly, somehow knowing where the fingertip is at all times. A systematic error cannot correct itself if no feedback exists. Many robotic arms operate in both modes, first without feedback for rapid deployment and then with feedback for fine adjustment.

As humans, we learn to perform many tasks and do many of them several times daily. As a result, we often forget how we learned a particular procedure and how much practice was required. To relive that learning experience, try using the "other hand" to punch in data in a calculator, or some similar task. The learning curve is sometimes very steep!

# 56. Farmer Chasing a Goose

If the farmer is restricted to chasing the goose along the instantaneous line of sight to the goose, the farmer will never be able to catch the goose unless there is a head-on encounter. The best strategy for the goose is to run in a straight line, for then the farmer's velocity is eventually in the same direction as the goose's direction, and the relative distance remains constant. Note that even when the goose changes direction often, the farmer cannot close the gap completely because the closer they are, the more toward being parallel are the velocities!

In the real open-field experience without the restriction, one strategy that might work if the goose is inexperienced is for the farmer to anticipate the position of the goose and to get there at the same time the goose arrives. However, most geese "read" the game plan and change course in midflight.

*Behroozi, F., and R. Gagnon. "The Goose Chase." American Journal of Physics 47, no. 3 (1979): 237–238.*

# 57. A Spooky Refrigerator

Yes. Just as you could remove a dot from a piece of paper with an eraser brought in from the third spatial dimension, a 4-D being could enter the refrigerator without needing to open the door and remove a piece of food. That is, 3-D objects are *open* in the direction of the fourth dimension.

Conceptually, visualizing a 4-D object is difficult in our 3-D world. Some mathematicians suggest letting the fourth coordinate direction be represented by flashing color, such as the sequence of colors in the visible spectrum from red to indigo. Take any 3-D object—a sphere, for example. As the sphere moves in the fourth coordinate direction its flashing color changes from red to orange to yellow, etc. A 2-D sheet of paper moving in the fourth dimension would be changing its flashing color also to indicate its fourth coordinate value. The inherent color of the object does not change, of course.

Descriptions of 4-D objects intersecting our 3-D world are quite fascinating. For example, a 4-D sphere intersecting our 3-D world would first appear as a point, then an increasing 3-D sphere, then a decreasing 3-D sphere, then a point, then gone! The analog in fewer dimensions would be a

3-D sphere intersecting a 2-D sheet of paper, being first a dot, then a widening circle followed by a narrowing circle, then a dot again, and then gone.

Although most people would expect there to be more mathematical difficulty and complications in even higher dimensions than four, this expectation is false. The mathematics actually simplifies with five dimensions and more! Much geometry remains to be worked out in a 4-D space, whereas the mathematics is better understood in the higher dimensions.

Gardner, M. The Colossal Book of Mathematics: Classic Puzzles, Paradoxes, and Problems. *New York: W. W. Norton, 2001, pp. 137–149.*

Peterson, I. The Mathematical Tourist. *New York: W. H. Freeman, 1988, pp. 82–107.*

Pickover, C. A. Surfing through Hyperspace: Understanding Higher Universes in Six Easy Lessons. *Oxford: Oxford University Press, 1999, pp. 44–70.*

# 58. Fractional Dimensions?

Yes. Noninteger dimensions are known as fractal dimensions. A pathway toward understanding fractal dimensions begins by considering duplications of well-known objects. A line segment can be duplicated to produce two line segments. A square can be duplicated in each direction to produce four squares. A cube can be duplicated in each of its three directions to produce eight cubes. In each case we obtain the number 2 raised to an integer power. We can make a table and generalize to d arbitrary dimensions.

| Figure | Dimension | No. of Copies |
|---|---|---|
| Line segment | 1 | $2 = 2^1$ |
| Square | 2 | $4 = 2^2$ |
| Cube | 3 | $8 = 2^3$ |
| Doubling figure | d | $n = 2^d$ |

We can now determine the dimension of an interesting but strange geometrical object, the Sierpinski triangle, named after the Polish mathematician who originally thought it up in 1916, shown here with its holes being gray. Double the length of the sides, and you get another Sierpinski triangle, similar to the first. For example, if the first Sierpinski triangle has one-inch sides, the doubled one has two-inch sides. How many copies of the original triangle do you have? Remember that the gray triangles are holes, so we can't count them.

Ignoring the hole in the center of the double-sized Sierpinski triangle,

we learn that doubling the sides of the original gives us three copies, so 3 = $2^d$, where d = the dimension according to the scheme in the table. Using a calculator, one finds that its dimension d = 1.585 . . . , a noninteger!

In general, the mathematical expression for the dimension of the figure is given by the ratio of two logarithms:

dimension = logarithm (number of self-similar pieces)/logarithm (magnification factor).

For simplicity:

1. A dimension between 0 and 1 is supposed to correspond to the capacity of a set of points to partly fill a line without achieving it completely, out of having the whole value 1 that is needed.

2. A dimension between 1 and 2 is supposed to correspond to the capacity of a line to partly fill a plane, without achieving it completely, out of having the whole value 2 that is needed.

3. A dimension between 2 and 3 is supposed to correspond to the capacity of a surface to partly fill a volume without achieving it completely, out of having the whole value 3 that is needed.

There is a whole world of mathematics to be learned with fractal dimensions and fractal geometry and their applications to the familiar physical world. One interesting question is whether two or more objects with the same fractal dimension must be related in some fundamental way, either mathematically or physically.

Peterson, I. The Mathematical Tourist. *New York: W. H. Freeman, 1988, pp. 114–142.*

# 59. Platonic Solids

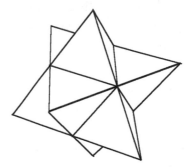

One must turn the second identical regular tetrahedron upside down and rotate by 30 degrees about the vertical axis before mathematically pushing them through each other to form a six-vertex regular solid. Obviously these six vertices correspond to the vertices of a regular octahedron if the new object were placed inside one. However, the sides are not convex and flat. Our bi-tetrahedron does have twofold symmetry axes even though the two tetrahedrons are rotated with respect to each other.

We might have asked you to place the two regular tetrahedrons face-to-face in congruence so that the combined object has five vertices defining a triangular bipyramid. There are now three twofold rotational symmetry axes, each one being through an edge of the joining faces. However, this bipyramid is not a regular polyhedron.

If we were to extend the discussion to four spatial dimensions, there are six regular solid convex objects analogous to the five Platonic solids in three dimensions. The number of regular convex solids, called regular polytopes, does not increase with more spatial dimensions. Instead, all dimensions beyond four have only three regular convex solids.

The Platonic solids are special in many ways, but perhaps their most important mathematical property was pointed out by mathematician B. Kostant:

> The ancient Greeks, especially the school of Plato, had great reverence for the regular polygons in the plane and regular solids in 3-space. The latter—the tetrahedron, cube, octahedron, dodecahedron, and the icosahedron—are often referred to as the Platonic solids. The Greeks believed that these regular figures were fundamental in the structure of the universe. If symmetry or its mathematical companion—group theory—is fundamental in the structure of the world, then one of the points of our lecture is the statement that the Greeks were absolutely right. That is, what we will be saying in a very profound way, the finite groups of symmetries in 3-space "see" the simple Lie groups (and hence literally Lie theory) in all dimensions.

One of us (F. P.) has proposed that the fundamental building blocks of matter, the leptons and quarks of the Standard Model of particle physics, are described mathematically by the specific rotational symmetries of the 3-D Platonic solids for the leptons and by their 4-D analogs for the quarks. The key arguments are that the mathematical symmetry groups for these regular solids are subgroups of the Standard Model symmetry group and that the lepton and quark mass ratios can be directly related to the ratios 1:108:1728 of the invariants of these subgroups. Whether the natural world mimics this fundamental mathematical behavior has yet to be determined by experiments at particle colliders.

Coxeter, H. S. M. Regular Polytopes. *New York: Dover, 1973, pp. 41–57, 126–144.*

Kostant, B. *"Asterisque." In Proceedings of the Conference "Homage to Elie Cartan," Lyons, July 1984, p. 13.*

Potter, F. *"Geometrical Basis for the Standard Model."* International Journal of Theoretical Physics *33 (1994): 279–305.*

# 60. Intersecting Spheres

Two identical three-spheres can intersect in a point, a circle, a sphere (two-sphere), and a three-sphere. Now bring in a third identical three-sphere to intersect with the former two in appropriate combinations of points, circles, spheres, and a three-sphere, the latter when all three are coincident. With three intersecting identical three-spheres, a resulting single two-sphere can be obtained only when the three three-spheres form a symmetrical configuration.

If the leptons and quarks of the Standard Model of particle physics are physical manifestations of the finite rotational symmetries of the 3-D Platonic solids and their 4-D analogs as proposed in a model by F. Potter (see the reference below), then the intersections of three-spheres will become important in fundamental physics. A quark would be defined in a 4-D space, and its mathematical behavior would depend on the properties of three-spheres. The proton, for example, is a real particle composed of three quarks in our 3-D world—that is, three 4-D entities according to the proposed model. So three three-spheres (representing the quarks) must intersect to form a two-sphere that "lives" in our 3-D space.

Potter, F. "Geometrical Basis for the Standard Model." International Journal of Theoretical Physics 33 (1994): 279–305.

# 61. Arm Contortions

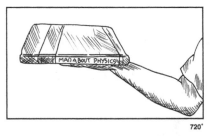

Yes, if one allows the arm to move overhead. This second rotation untwists the arm and brings the orientation of the book back to the initial one again. One can say that the arm-object pair requires two 360-degree rotations to return to the initial orientation. Such an entity is said to mathematically correspond to a spin ½ system—that is, related to the continuous symmetry group SU(2). Any lepton or quark wave function, such as the electron wave function, behaves in this way with respect to rotations and angular momentum.

Spheres, cubes, and other objects with spatial symmetry also can be classified as spin ½—that is, their rotations are described by symmetry groups that are subgroups of SU(2) and SU′ (2) = SU(2) × $C_i$, where $C_i$ is the two-element inversion group. For the Platonic solids, the rotational symmetry groups are discrete instead of being continuous, and some of these symmetry groups are subgroups of both SU(2) and SU′ (2) because among all the elements of both can be found the elements of finite order for the discrete subgroups.

As you know, our practical experience is mostly with spin 1 entities—that is, those needing a 360-degree rotation to return to the initial orientation. Mathematically, these spin 1 properties can be constructed from the spin ½ symmetry properties. Mathematicians know that even more fundamental are the reflection groups from which all spin ½ properties can be derived as two reflections in perpendicular planes. The two books listed below discuss these hierarchical relationships and many more.

Altmann, S. L. Rotations, Quaternions, and Double Groups. *Oxford: Clarendon Press, 1986.*

Coxeter, H. S. M. Regular Polytopes. *New York: Dover, 1973.*

Rieflin, E. "*Some Mechanisms Related to Dirac Strings.*" American Journal of Physics 47 (1979): 379–381.

# 62. The Rotating Cup

Done properly, you would see the same sequence in both cases. The cup appears to rotate with changing rotation rates as time passes. We have here an example of Galilean relativity for uniform motion. At these familiar slow speeds compared to the speed of light, the behavior of the rotating cup produces no surprises. We can walk past the cup, or the cup can move past our stationary location.

In a later chapter, where we introduce the special theory of relativity (STR), we examine an object such as a cube moving past a stationary observer at enormous speed, and we could consider the opposite case, of the observer moving past the stationary object. Of course, one sees the same behavior in both cases, just like the symmetry we observe in Galilean relativity. However, in STR an effect, now called the Terrell effect, explains why a cup approaching and passing by at near-light speeds appears additionally rotated, the observer being able to see the back side of the cup as it approaches!

# 63. Space and Time Together

Using three real spatial coordinates and one imaginary time coordinate for calculations works correctly when calculating the squares of space-time coordinates and their sums and differences. The important relationship is the space-time interval $\tau$ defined by $\tau^2 = c^2 \Delta t^2 - \Delta x^2 - \Delta y^2 - \Delta z^2$, where the $\Delta x$'s are the four "distances." However, physics textbooks that use an interval $\tau$ defined by $\tau^2 = + \Delta x^2 + \Delta y^2 + \Delta z^2 - c^2 \Delta t^2$ are making a mathematical faux pas in choosing three real space coordinates and one imaginary time coordinate—that is, the set $(x, y, z, ict)$ with i being the imaginary and c being the speed of light—instead of vice versa. Fortunately, this fundamental error does not affect the calculations of time intervals and spatial separations because these calculations involve the differences of squared quantities. To be mathematically correct in the $(3 + 1)$-dimensional space-time, one must use quaternions, which are numbers in the form $q = a + bi + cj + dk$, with $i$, $j$, and $k$ being $\sqrt{-1}$ and $a$, $b$, $c$, and $d$ being ordinary real numbers, because they are the numbers in four dimensions that properly handle rotations, translations, and Lorentz transformations. Today, quaternions are used everywhere in science to describe the dynamics of motion in three-space. Spinors are equivalent mathematical entities used in quantum mechanical wave functions to describe the electron and other fermions in $(3 + 1)$-D space.

Quaternions were first "discovered" by W. R. Hamilton in the 1800s, and the quaternion $q$ has one real component and three imaginary components. Just as complex numbers are formed from pairs of real numbers, quaternions are formed from pairs of complex numbers. Thus one should assign the time coordinate to the real component and the three space coordinates to the three imaginary components of a quaternion. Hence, mathematically, we live in a quaternion world with an imaginary 3-D physical space and a 1-D real-time clock!

Altmann, S. L. Rotations, Quaternions, and Double Groups. *Oxford: Clarendon Press, 1986.*

Pickover, C. A. Surfing through Hyperspace: Understanding Higher Universes in Six Easy Lessons. *Oxford: Oxford University Press, 1999, appendix D.*

# 64. Space > 3-D?

There are several arguments for why space is not larger than three dimensions. Planetary orbits are not stable when n > 3, except for a circular orbit for n = 4, because the attractive force and the centripetal force both do not have the correct dependence on radial distance. In 1917 P. Ehrenfest showed that one needs to consider the Poisson equation for arbitrary dimensions to determine orbit stability. When the $n \geq 4$ circular orbit for a body around a central mass becomes slightly perturbed, one can show that the comparison of the central force to the centripetal force for the orbit depends on the perihelion value $r_1$ and the aphelion value $r_2$ according to $[1/2 - (n-2)^{-1}]/r_1^2 < [1/2 - (n-2)^{-1}]/r_2^2$, which cannot be true for $n = 4$ and larger. In a 4-D space, a satellite launched from Earth toward the Sun would either fly away to infinity or spiral into the Sun.

The hydrogen atom is not stable when $n > 3$ because there is no energy minimum for $n \geq 5$, which is shown using the indeterminancy principle—that is, the Heisenberg uncertainty principle. For the case $n = 4$, the relativistic energy equation must be examined to show that no energy minimum is available and the atom is not stable.

Several other physical phenomena would be unusual for $n \geq 4$ dimensions. There is no satisfactory propagation of sound waves or electromagnetic waves free of distortion and reverberation in spaces other than $n = 1$ and $n = 3$. Also, axial vectors such as the magnetic field and the angular momentum vectors do not exist in even-dimensional spaces.

The considerations could be extended to a universe with more than one time dimension! However, this matter and others we leave for future challenges.

Büchel, W. "Why Is Space Three-Dimensional?" Translated by I. M. Freeman from Physikalische Blätter 19, no. 12 (1963): 547–549. American Journal of Physics 37 (1969): 1222–1224.

Ehrenfest, P. Annalen der Physik 61 (1920): 440.

Pickover, C. A. Surfing through Hyperspace: Understanding Higher Universes in Six Easy Lessons. Oxford: Oxford University Press, 1999, pp. 202–205.

# Chapter 4
# Fly Me To The Moon

## 65. Gunfight

One could do good classical physics here, but the filmmakers have turned the scene into Hollywood exaggeration. The physics is determined by the conservation of linear momentum. Assume that the victim of the shooting is initially at rest, so the total momentum initially is all in the bullet (or buckshot) of mass $m$ and speed $V$ before hitting the victim, at $mV$. After the collision, the final total momentum is in the backward "flying" person plus bullet. If the victim has mass $M$ and the combined victim-plus-bullet object has speed $v$, the total final momentum is $(M + m)v$. For all interactions, by the law of conservation of linear momentum, the final momentum equals the initial momentum.

In the simplest case (assuming no frictional drag at the feet and ignoring transfer of momentum away to the earth, etc.), its application yields $(M + m)v = mV$. Solving for the velocity $v$ of the victim afterward produces $v = mV/(M + m)$. Substituting reasonable values of $M = 80$ kg, $V = 400$ m/s, and $m = 0.03$ kg yields a "blow-back" velocity maximum of $v = 0.15$ m/s. Most people can walk about 2 m/s (i.e., about 4 mph). So we can conclude that any shooting victim depicted as being blown backward by the impact of the bullet (or shotgun blast of pellets) is ridiculous and belongs in the fantasy world only!

A physicist wouldn't actually need to calculate the velocity backward using linear momentum conservation explicitly. Simply watching the behavior and movement of the shooter holding the gun before and after the shot reveals approximately how much momentum is available by using Newton's third law. If the shooter isn't blown backwards by the recoil force of the shot, the victim won't be either. Of course, someone will suggest that involuntary muscle contraction in the stunned victim causes the "flying"

backward. Falling backward, perhaps, but not "flying"!

There is the story of a famous physicist back in the 1950s who loved to watch gunfights in Western movies. The bad guy always draws first, he noticed, but the good guy wins the gunfight. How could this outcome happen? His hypothesis was that psychology played an important role, slightly hindering the man who had to make the conscious decision to draw first. The second man simply had to react.

Even today, the psychology of choosing a physical action is an important factor, particularly in sports. There are tennis coaches (and coaches in other sports) who preach the psychology of playing tennis, saying that when you think too much on the court instead of simply reacting, as you learn to do in practice, then you are in trouble. You are letting self no. 1 (your mind) control self no. 2 (your body), and your tennis game will suffer. We wonder whether the concepts hold true for playing the physics game, too!

## 66. Body Cushion

We doubt whether landing on top of another body after such a long fall provides much cushion! The important parameter here is the extent of the collision time $\Delta t$—that is, how long the collision of the hero's body with the object actually lasts. The longer the $\Delta t$, the better. We also need to know the acceleration $a$ versus time profile. In better words, what is the maximum acceleration to be experienced by the hero's body? By definition, the average $a = \Delta v/\Delta t$, where $\Delta v$ is the velocity change during the time interval $\Delta t$. Shorter $\Delta t$'s make the experience more painful.

Stunt professionals are often seen leaping off buildings or falling through windows in movies, but their collisions are with huge air-filled balloons that effectively extend the total collision time to half a second or more. We do not see their collision with the balloon in the program because the editing process substitutes the desired body lying dead on the concrete.

Back to the hero landing on the other body. The collision time here will be less than one-tenth of a second, producing dangerous accelerations. For example, if the body falls from the top of a two-story building, its speed will be approximately 11 m/sec just before collision. The acceleration during the collision will be greater than 110 m/s$^2$, very dangerous. Even if bones are not broken in bringing the hero's body to rest, the internal organs will continue to move until they suffer a collision inside the hero's body. And if the hero bounces back upward, the acceleration can be even worse,

because the change in linear momentum will be almost double, even though the collision duration may be increased slightly. Automobile collisions provide plenty of evidence about the damages done to internal organs by sudden collisions with very short collision times. We doubt whether our hero will be able to walk away from the "body cushion." In fact, our hero will be lucky to survive!

# 67. Cartoon Free Fall

When the cartoon character steps off the cliff, the fall should begin immediately, of course. The natural path is essentially a parabola, with approximately free-fall acceleration downward and a constant velocity horizontally. Even a cartoon character must have some mass; otherwise the character could not exert a force on anything, including the ground being walked on. Unless the upward buoyant force of the air balances exactly the gravitational force downward, no cartoon character stepping off the cliff would remain in suspension at the height of the cliff. Even if the buoyant force was sufficient, why would its upward push disappear suddenly to allow the character to free fall?

We should see the character accelerating downward with ever-increasing speed unless the terminal velocity is reached or the buoyant force balances the weight. The collision at the bottom is also subject to analysis. To prevent this sudden collision, sometimes another character is able to run down from the clifftop just in time to catch the falling character. And sometimes we even see another person falling with a greater acceleration downward to arrive in time to catch the victim! If the fall is nearly at the free-fall acceleration, the runner must be mighty swift! There are measured examples of skiers going down Mount Fuji with accelerations greater than the free-fall acceleration, but no runner has achieved this feat yet. And yes, an anvil always has a greater acceleration downward than any other object (sure!)!

# 68. Silhouette of Passage

The condensed matter physicist knows a lot about the physical properties of liquids and solids, so he or she probably would say, "Wow! How was *that* done?" The only realistic possibility is that the wall cut should be quite messy and the cartoonist cleaned the edges for heightened dramatic effect!

We can estimate how difficult the cookie cutter hole would be to achieve by considering a ball thrown at the wall. The impact surface area increases rapidly in time as the ball and wall both bend a little during the collision. The initial kinetic energy of the ball just before the collision becomes distributed in the distortions of the ball and the wall. The interactions between the molecules of the wall material change as some of the available energy from the collision spreads from the immediate impact area. If the total collision time is extremely short, the energy distribution will be quite limited in distance, and much of the energy is available for ripping. Otherwise, if the collision time is much longer, a big portion of the wall will respond by deforming just a little, and ripping may not occur.

A bullet going through the paper at a practice target makes a fairly clean hole for two reasons: (1) the contact time is extremely short; (2) the bullet offers a nearly round profile, so there is symmetry about an axis perpendicular to the hole. Even then, close examination of the bullet hole reveals an irregular surface and additional tearing beyond the actual round hole.

You can check out the advantage of a symmetrical object. Now, with appropriate safety precautions, try to rapidly push any shaped profile other than round through a sheet of target paper. The lack of cylindrical symmetry perpendicular to the surface usually creates enormous problems for the material because slightly more energy goes into some directions. Moreover, tearing is required to occur at different distances along a noncircular profile, so there may be points where the transferred energy density is significantly higher or lower than the surrounding paper regions. All these factors and several more act against a very clean cut through the material. The cut through a thicker piece of paper or a wall that has significant depth would be even messier.

Of course, the cartoon character has several other options for getting through the wall when time permits: (1) simply paint an exit onto the wall through which only he can pass, or (2) the character can carry around a hole to be affixed where needed!

# 69. Artificial Gravity

Space stations and spaceships have been created by authors and screenwriters in a vast array of shapes and sizes. A rotating dumbbell shape has appeared in many space adventures, the rotation providing a pseudo-gravity force for Earth creatures. The rotation about the center of mass perpendicular to the long axis provides a pseudo-acceleration acting radially outward called the centrifugal acceleration $a_c = v^2/r$, where $v$ is the tangential velocity value and $r$ is the radial distance from the axis of rotation. The resulting centrifugal force is called a pseudo-force because the actual force is acting radially inward toward the axis of rotation to accelerate the object from its inertial straight-line motion. We must assume that the structural integrity of the space station remains intact—that is, the station was designed for the rotation and for the allowed distributions of mass on board.

Using the relation for the angular velocity $\omega = v/r$, we can express the centrifugal force as $F_c = mr\omega^2$. The whole spaceship has the same angular velocity about the rotation axis, so an object's radial acceleration increases linearly with distance $r$ from the rotation axis. An astronaut at one end of the dumbbell must climb—that is, walk up a ladder and then down a ladder—from one end to the other through the middle, where the radial acceleration is zero. The muscular effort required changes throughout the climb, so the sensations must be wonderful!

# 70. Small Wings

The small wings on such alien beings are probably much too small for a 20 kg body. One could argue that the planet's gravitational force at its surface is much less than the value here on Earth, so the alien being's weight is much less also. That may be so, and the proposition is not unreasonable. However, we still require a sufficient air density for the wings to do their work and a breathing atmosphere for our Earthling on the foreign planet's surface. (After all, this example has the human standing there breathing without any special oxygen supply.)

We need to determine the required density of the atmosphere of this alien planet, assuming an adequate supply of oxygen molecules for breathing by our visiting human. Earth's atmosphere at sea level has a total density of about 1.4 kg/m³, of which $O_2$ comprises about 20 percent by molecular composition. That is, 1 cubic meter of air weighs 1.4 kilograms. The remainder, of about 80 percent, is $N_2$, which has a molecular weight of 28, compared to 32 for $O_2$. For simplicity, we assume they have the same molecular weight, so we require an alien atmosphere to have about 0.3 kg/m³ of oxygen available for breathing.

The gravitational acceleration $g'$ at its surface determines the air density at the surface for a given molecular composition and air temperature profile. Most planets will have an acceleration not much different from the value of 9.8 m/s² here on Earth, as one can check out for the planets in the Solar System, for example. So the wings must be capable of exerting an upward force at least as great as the downward gravitational force—in our example,

the weight $F = g'\, m$ of the alien being with small wings, or 200 $N$ if $g' = $ 10 m/s².

We assume that a very strong 20 kg individual can stretch out his arms horizontally to the side and push upward against two supports with about 200 $N$ downward force to lift the body. However, this same individual will not be able to use small wings of the same length and perhaps just a little wider than the arms to beat against the air with equal effect. If you doubt this hypothesis, put some arm-length wings on a strong person and observe how easily he or she can lift off and hover a few centimeters above the ground!

If the alien being were hollow inside so that its total mass is significantly less than expected for the body size, there may be no problem with hovering.

## 71. Shrunken People

We assume that the proposed shrinking to one-hundredth scale can be accomplished. Unfortunately, your weight would remain the same (unless you get rid of molecules somehow), and your density would increase a millionfold! The area of contact of your feet would be 10,000 times smaller, so the pressure at your soles would be 10,000 times greater, rising to about 20,000 psi. Every step would break the concrete, or you would sink into the ground until the upward normal force could balance you. Among other changes, your metabolism must change enormously, for your high ratio of surface area to volume will mean that the rate of heat loss has increased at least 100 times. Of course, we choose to ignore any consequences inside the body for simplicity.

Notice that if the opposite happens and you grow bigger and increase your size by a factor of 100 in all directions, without adding molecules, your density decreases a millionfold. You would be blown away by practically any breeze! But your greater problem would be that your density is now much less than the density of air, so the buoyant force upward would be greater than your weight. You are now a giant balloon being pushed upward toward the upper atmosphere! Also, your metabolism would change dramatically, but again, we ignore any consequences inside your body. You could probably make the journey around the world in 80 days without the hot air!

## 72. Spaceship Designs

Landing a spaceship on Earth and then taking off for space involve the same forces, but the gravitational force always acts toward the center of Earth, sometimes being a help and sometimes being a hindrance. The major problem is the enormous energy requirement in getting from the surface of Earth to a reasonable distance away. Once the spaceship is more than a few Earth diameters away, its nuclear engine operation can be reasonably efficient in accelerating the vehicle. However, to get off the surface requires a tremendous amount of energy, *and* its rocket engines must throw out a lot of momentum in the exhaust gases at high speeds to achieve "escape velocity." Newton's third law dictates this momentum requirement. The particles ejected backward act on the rocket in a force pair, the rocket pushing particles backward while the particles are pushing the rocket forward.

To reach outer space from Earth, the vehicle must provide a large supply of energy and be able to eject a large amount of momentum, usually by having a large supply of mass to eject. The energy needs can be accommodated by a variety of engineering designs. However, the physics is quite demanding on the amount of mass ejected per second. The fuel mass used for this propulsion is not consumed instantly, so this fuel mass adds to the mass of the vehicle at launch time. Consequently, even more propulsion fuel mass and energy are required for a launch than simply accounting for the payload itself. The mass of the fuel supply soon becomes many times larger than the actual payload launched into space.

So when the spaceship leaves its Earth spaceport and doesn't eject a lot of stuff backward out of its rocket engines, the film is expressing a mode of operation that is not achievable with present technology. But perhaps the propulsion will be different in the future, say the disbelievers. So let's now go to the extreme propulsion limit. The most efficient process would be particle-antiparticle annihilation, conversion of fuel and antifuel completely to energy in the form of high-energy photons according to Einstein's famous $E_0 = mc^2$. If we ignore many problems such as a source for antiparticles, radiation exposure, and so on, and also assume that all the photons are eventually directed rearward, each kilogram of fuel could provide $3 \times 10^{16}$ joules of energy and $3 \times 10^8$ kg-m/sec of linear momentum. To accelerate upward at about 10 m/s², a kilogram of this fuel can provide a million-kilogram spaceship with 30 seconds of thrust. If one requires 3,000 seconds of thrust, simply use 100 kg of matter-antimatter fuel. We look forward to the future of space travel with antimatter engines, but for the present we can enjoy the entertainment provided by space travelers in films.

# 73. Warp Speed

Spaceships with a warp drive to accelerate beyond the *local* speed of light cannot be ruled out just yet! One example is the expansion of the universe, which carries everything along, and speeds can exceed the speed of light. Distant quasars with recession velocities greater than c are a reality. The analog on Earth may be to have a 100-meter race on a stretchable track that can change length during the race.

However, if the spaceship has the technology to be able to distort space-time itself in its local vicinity, then there is no need for enormous speeds. Simply contract the space in the front to bring distant points closer, warping space-time directly. That starbase that formerly was light-years ahead of you is now close to you, reachable by normal propulsion in minutes!

Krauss, L. The Physics of Star Trek. *New York: HarperPerennial, 1996, pp. 56–58.*

# 74. North Pole Ice Melt

There would be no change in the sea level if all the ice at the North Pole melted. Why? Because this ice is floating on water. Upon melting, the water molecules in the ice simply occupy the space of the liquid displaced by the ice originally. Of course, we need to be more careful on defining the extent of the North Pole. If we include ice on some landmass, then this ice will add extra water molecules to the liquid seas and slightly raise the sea level. In contrast, most of the ice at the South Pole is several kilometers thick and is predominantly on the Antarctica landmass, so its melting could significantly raise the sea level. Some films depict oceans that have risen hundreds of meters from the worldwide ice melt. Simple estimates easily reveal that this conjectured amount of sea level change is ridiculous.

Another concern might be the linear expansion of water when its temperature is above 4°C of about $70 \times 10^{-6}$/°C. Even if the water temperature rose by 10°C throughout the first 10 kilometers of ocean depth, the expected water level rise for a column of water would be no more than 7 meters if the surface area remained constant. But the surface area will expand, so the actual water level will rise about 2 meters of so. For a 1°C increase in temperature, the ocean level rise will be several centimeters.

Sea level changes played major roles in the migrations of our human ancestors. Some Bushmen left their homeland and their cave dwellings about 40,000 years ago, seeking better climates and less arid lands. A mini ice age had developed, so seawater became locked into the ice at the poles. Their caves, which were originally near the sea, were now several hundred kilometers inland from the sea, and the arid climate made living off the land very difficult. So they migrated along the eastern coast of Africa all the way through the Middle East and India to Australia. The Aborigines in Australia are descended directly from these peoples from Namibia and South Africa and together with the Bushmen are the oldest civilizations on Earth.

# 75. Lightning and Thunder

Let's consider identical explosions on the battlefield at two different distances from the observer (i.e., the camera) but seen simultaneously. The sound intensity emanating from the farther explosion should be less loud compared to the closer one, and the sound of the farther explosion should be delayed more with respect to its light flash than for the closer explosion. The extra distance affects both the sound intensity and its time delay. Depending on the distance and the temperature gradient in the air, there could be additional effects, such as different frequencies having slightly different speeds and/or paths en route.

For example, if the closer explosion is one-fifth of a mile away and the farther one is two-fifths of a mile away, the differences in arrival times should be clearly heard. The sound from the closer one begins to be heard *one whole second* after the light flash and the weaker sound from the farther explosion should be heard one second later, *two seconds* after the apparently simultaneous light flashes. Count these time delays aloud and you will immediately realize that most battlefield scenes do a false portrayal of the audio timing. But who really cares? The dramatic license enhances our viewing experience of the fantasy world of the cinema! But experienced military people know the difference.

Many more audio violations can be recalled from the movies. We simultaneously hear and see airplane crashes in the distance, cars falling and crashing into a chasm far below, jet airplanes passing overhead with no sound delay, etc., all for the benefit of a moviegoing public who should know the difference.

# 76. Explosions in Outer Space

The colors of the explosion are probably okay, and perhaps one can appreciate the beautiful outward-going streamers isotropically distributed. However, very seldom does a real explosion, even in a vacuum, distribute the stuff isotropically. One also would expect bits and chunks of vastly different sizes, with a large chunk or two left over near the explosion origin.

As a dramatic example, in 1987 Supernova 1987A in the Large Magellanic Cloud, bound to our Milky Way galaxy, blew apart, dumping practically all its energy into neutrinos, but there was still enough energy to produce a lot of photons for a bright flash of visible light that was first seen by amateur astronomers in Japan. Today that light flash continues to expand with decreasing intensity, and the gas cloud of particles also continues to stream outward, impacting molecules and other stuff in various directions to help "paint" the region in beautiful colors. Data seem to indicate that the remains at the origin consist of two small, massive objects orbiting a common barycenter.

The real problem with hearing an explosion in space is that there is no medium to carry the sound waves, so no sound from the original explosion should be heard by the spaceship crew or by the audience in the theater. The light travels at nearly $3 \times 10^8$ m/s, and the sound would transit at the snail's pace of $3 \times 10^4$ m/s or slower. The light flash comes before the sound for any safe distance. Of course, any debris from the explosion hitting the spacecraft would make an impact sound carried by the walls of the ship and by the internal artificial atmosphere. Then there is the noise of the other spaceship going past, but that is another story.

## 77. Space Wars

Don't try to learn your physics from space battles. Practically everything is incorrect, except the ability to cause something to explode if your weapon can dump enough total energy into the target in a short time!

The laser beams, no matter how powerful they are, would not be seen en route. Seeing them requires that a reasonable amount of light be scattered back to you along their paths. But there is practically *nothing* in the vacuum of space to scatter off. A few hydrogen atoms per cubic meter are all there is out there.

The explosion would not be heard at all. Sound requires a medium for its transport, and the vacuum of space is not a material medium able to conduct sound waves. You will feel something, the impact of the bits of the exploding debris from the enemy battlecruiser because they travel unimpeded from the explosion volume to your spaceship. Their speeds may be enormous, so they could do significant damage if your shields are not in place.

## 78. Security Lasers

As the director of the theft scene, you would know that the laser light would not be visible in the air in a normal room because there is not enough scattering of the light to your eyes by the molecules in the air. Spots where the beams strike walls, mirrors, or any objects could be seen, but their straight paths to the spots are invisible. Therefore you have two ways to make the laser beams visible so your audience can see the sequence of maneuvers required to succeed in the heist: (1) put something in the air itself to scatter laser light such as chalk dust, smoke, or a fog of liquid nitrogen droplets or water mist, or (2) artificially put in the light beams during editing, after the scene is done.

We suppose that some real-world locations may use an arrangement of crisscrossed laser beams for security aids, but we do not know of any. Certainly, infrared lasers are much cheaper than visible lasers, and they would not be seen by injecting a fog into the space. Nor would their spots on the wall be seen.

Some heist scenes show the substitution of a mirror to fool the security system while the protagonist steals the item. Indeed, the mirror substitution may reflect the beam in the correct direction, but the tenths of a second disturbance in the original beam would be very easy to detect by the security system electronics, which can sense disturbance spikes quickly and sound an alert. Of course, the human security operator may choose to ignore the alert, which so often happens in movies!

## 79. Bullet Fireworks

Normal bullets are copper-clad lead and do not spark upon impact with steel or any other surface. Just go to a grinder to check out the properties of copper versus other metals in the production of visible sparks. The grinding of steel produces sparks everywhere, seen even in sunlight. The many small, hot particles of steel are actually burning. Now grind a piece of copper tubing. No sparks. You might see an occasional spark due to contamination on the grinding wheel or in the copper. The copper bits ground off do react with oxygen, but they do not get very warm. Please do not grind lead because toxic particles will be released into the air and, besides, lead is known not to produce sparks. So the conclusion is that practically no bullets produce a brilliant flash of light on impact.

In support of the film depictions is the fact that the military does have machine gun bullets containing white phosphorus so that the point of impact can be seen by the gunner. These bullets are used also to ignite fuel tanks and other possible containers of explosives by producing sparks to ignite the vapors. But phosphorous bullets are very rare outside the military.

## 80. Internet Gaming

The Internet is usually not the culprit, being extremely fast compared to the hand-eye coordination of the player. The actual travel time between the major switching stations along the Internet is extremely short because the data packages travel at nearly the speed of light in a vacuum. So a data package can go 20,000 kilometers in about 70 milliseconds with no delays. Delays at the switching stations along the Internet are typically in hundreds of milliseconds, with your local Internet service provider (ISP) contributing most of the delay time, on the order of 300 milliseconds or so.

In contrast, your local computer and cable modem, for example, usually will be slower in responding to your input and getting the data out to your ISP and onto the Internet. The faster the equipment at your end, up to a certain speediness, the quicker your response will appear in the game. If your equipment delay time becomes less than the Internet delays, there's nothing further to be improved by spending more money on a faster computer system.

## 81. Cartoon Stretching

There are many ways to approach the problem of determining the speed of sound in the cartoon character's body. We consider one approach only. Start with the application of an external force to the surface of the body at some location—the character's foot, say. We see the stretch region progress up the leg over a period of one to two seconds.

"Wait just one minute!" you exclaim. How is the speed of sound in the material related to its speed of stretching in response to an applied force? The answer is that both processes require communication from one molecule to the next molecule outward, from the applied force region to the far reaches. Usually the much smaller energy in the sound application produces a very tiny stretch followed by a relaxation and overshoot, then another stretch, etc., repeatedly at some frequency above 14 Hz or so. The stretch produced by the tug of a much larger applied force shows a much larger displacement of the molecules that may or may not relax when the applied force is lessened. The larger displacement between molecules for the stretch process may require slightly more time if the process cannot be modeled by a collection of linear harmonic oscillators, but the speed of stretch will be very close to the value of the speed of sound for most materials.

The stretched cartoon character's body exhibits a communication speed of about a meter per second, a very slow speed of sound indeed when compared to most materials, which have a speed of sound of about 300 meters per second. So we conclude that cartoon characters are made of very unusual materials. Perhaps designer materials in the future will be able to mimic a cartoon material. According to the world's greatest detective, the game is afoot!

# 82. Infrared Images

The actual converted infrared image would be a *blurred* black-and-white image, not a sharp one. The physics places a limit on the resolution of images we see in the infrared compared to the visible. When we look through binoculars or any lens system in the visible, these systems have a resolution limit that depends on the quality of the optic elements. No matter how much the image is enlarged or enhanced by dithering, etc., the original resolution is not improved even though the image looks cleaner. But the physics is even more restrictive when comparing an infrared image to a visible image because the visible light has been coherently scattered from the object, whereas the infrared light has not, as explained below. Of course, one cannot exceed the Rayleigh criterion for resolution of approximately one wavelength of the light, except by using interference techniques, which we do not consider here.

In the visible part of the electromagnetic spectrum, atoms absorb and emit the photons of light in a two-step process, usually absorbing and emitting in about $10^{-16}$ second. During this time interval, the molecule holding that atom moves very little. Nearby atoms also scattering this impinging beam of numerous photons tend to remain in place during their scatterings. In effect, during the scattering of each photon there always will be a fixed phase relationship among all the atoms scattering light from the object to your light sensors. Hence, photons scattering from different areas on the object's surface carry detailed phase information with fixed phases to achieve nearly maximum resolution. If the phases actually varied randomly from one location to another, the visible image would become blurred.

In the infrared, the image is blurred because most of the infrared is absorbed and emitted by molecular vibrations and rotations that have random phases over the object's surface. This scattering process takes much longer—about $10^{-12}$ second—sufficient time for the molecule to move considerably during the scattering, so there will not be a fixed phase relationship among neighboring molecules on the surface of the object. The same surface that appeared well resolved in the visible will now be quite blurred in the infrared. No magical digital techniques will be able to take an infrared image and make a sharpened black-and-white image true to the original object.

In the ultraviolet, the scattering time is very short, but so is the wavelength, so the extent of the coherent scattering area also is very short; thus the image is blurred compared to the visible one. Nature has given us a vision range in the visible that ensures the best resolution of detail.

# 83. Light Sabers

Yes, the light sabers would pass through one another as if nothing were there! The photons of the light are bosons, which do not repel each other. The clashing of the light sabers is pure artifact and far beyond artistic license. It's an outright lie to the public!

There are two general categories of particles in the universe: fermions and bosons. Two identical fermions (e.g., think electrons, or any other fundamental spin 1/2 particle) cannot exist in the same quantum state defined by its 4-momentum and spin. The existence of all matter, including us humans, critically depends on the inability of two identical fermions to get together in the same state, so matter occupies a volume. That is, objects can be bigger than a point!

In the case of bosons, one can not only put as many identical bosons (think photons or any integer spin fundamental particles) into the same quantum state, they also prefer to be in one, with the probability to do so enhanced by the number $N$ of bosons already in the state. There is no repulsion experienced. So two light sabers intersecting at an angle would pass right through one another with no change in either. If the light beams were powerful enough, though, they could cut material objects made out of fermions—that is, the ordinary stuff all around us—whenever this material stuff intercepts the beams because sufficient energy could be absorbed to change the physical state of the material and result in vaporization.

## 84. Force Fields

We don't know why we can see the good guys through the battlefield force field via visible light at the same time that the visible laser beams cannot get through! This result might be due to some dramatic nonlinear natural effect not yet experienced in research labs, or the phenomenon is a pure artistic falsehood. We would bet on the latter. Obviously, the light transmission properties of a transparent material can change dramatically on absorbing energy, but the usual effect is hole-burning, not reflection.

Then there is the problem of why the laser beams can be seen along their paths, unless the dust in the air above the battlefield scatters enough of the light. But when the laser beam originates in space where there is no air (or other scatterers in significant density), one can still see the laser beams. Is there no correct laser physics in these movies? Is there any correct physics at all?

## 85. Cold Silence of Space

Yes and no. The very good vacuum between Earth and Venus, for example, certainly does not conduct sound, so the space is silent. But what is the temperature of outer space? This question is an improper question, for we must instead ask: "What temperature would be recorded by a thermometer placed in space between the Venus and Earth orbital distances?" Whatever kind of thermometer we use, if the sensor is not rotated, one side will be exposed always to the direct rays from the Sun and the other side will be in shadow. At thermal equilibrium, the amounts of outgoing radiation energy and incoming radiation energy in all directions will balance.

The equilibrium temperature $T$ for an object at Earth's average distance from the Sun is about 280 K, or $+7°C$, the actual value being lower because there will be some energy reflected away and not absorbed by the thermometer. One calculates $T$ from this equation: flux absorbed = flux emitted. If we assume for simplicity that the sensor is a sphere of radius $R$, then the equation becomes $S(1-A)\pi R^2 = \sigma T^4 (4\pi R^2)$, where $S = 1.4$ kW s$^{-1}$ is the solar constant at Earth distance and $\sigma = 5.67 \times 10^{-8}$ W m$^{-2}$ K$^{-4}$. The parameter $A$ is the reflectivity, which we take as zero in the ideal case of a perfect absorber and radiator. Real materials will have an A value between 0 and 1.

If you are orbiting at the Earth distance from the Sun, you may desire to rotate your spacecraft slowly in space so that all sides cook evenly! This passive heating can be augmented by active heating from within to maintain a cozy environment.

If you are orbiting closer to the Sun than Earth, the equilibrium temperature will be higher, the solar flux increasing as the ratio of the squared distances. Near the orbit of Mercury you may be too hot! If you are farther away, the temperature decreases, so you may need artificial heating. Some rotation may produce a system that requires less fuel for heating, but the details need to be worked out.

# 86. Nuclear Submarine

Submariners love to dive in their submarines. A dive to several hundred meters under the surface *may* be effective in limiting the initial spread of the debris from the *thermal* explosion. No nuclear explosion would occur, or else everything around would be vaporized by the energy released, in which case the depth in the water would help very little in preventing the spread of energy in many forms. The thermal explosion in a nuclear reactor in the sub releases the fuel and coolant, so radioactive particles and debris will be sent out in all directions. Some of this stuff would be slowed effectively by the water, and some probably would make the surface and escape into the air.

When the Chernobyl nuclear reactor broke its containment vessel in the 1980s, its nuclear particles were detected around the world within hours to days. At the University of California at Irvine, the air filtering system at the local nuclear reactor recorded the radioactive cesium and iodine particles in parts per billion from the Chernobyl incident 10 days after the chemical explosion.

# 87. Plutonium vs. Uranium

A plutonium bomb would be much safer to handle. Weapons-grade plutonium (Pu-239) emits primarily alpha particles and low-energy gamma rays, both being easy to shield. The trace amounts of even-numbered Pu isotopes have spontaneous fission reactions that emit neutrons. However, neutron detectors would need to be within a few meters to detect these neutrons above background. So a lost plutonium bomb could be very difficult to find.

The plutonium usually is coated with beryllium or another appropriate sealant because the exposed element will react chemically with the oxygen in the air or in water and increase its temperature considerably. The person holding a plutonium bomb would probably feel that the protective casing is warm, because the alpha particles deposit their kinetic energy in the casing material.

Of course, inhaled or ingested Pu is one of the worst carcinogens known. Any explosion releasing Pu into the air creates a hazard for all life that would remain for a long time.

In contrast, weapons-grade U-235 releases gamma rays at several energies, the most intense at 186 KeV. So the detection of a device containing U-235 is much easier than trying to find plutonium. Some films portray the differences correctly, while other films dramatize any nuclear device and its possible dangers with remarkably adolescent scare techniques.

# 88. Nuclear Detonation

We know of no nuclear explosive device that does not require at least very good spherical symmetry to be detonated. The simpler atomic devices have either two hemispheres that must be rapidly moved together into a sphere, or two spherical sections held apart until a slab of nuclear material is shot into the gap to start the fission reaction. The more complicated hydrogen devices require a strong, spherically symmetrical implosion from the perimeter shell to initiate the fusion reaction.

Dropping the weapon from any height would damage the casing asymmetrically. Even shooting pellets or bullets, etc., through the casing into the warhead would create an asymmetrical result but no explosion. Initiating the nuclear reaction and keeping the reaction going are not easy. The process is certainly not worth worrying about to the extent portrayed in the movies. One should be more concerned about whether the proper safety precautions are being practiced against accidentally dropping the thing on one's toes.

By the way, the smallest practical nuclear weapon tips the scales at only about 9 kilograms, about 20 pounds, and is small enough to fit into a bulky attaché case. For the physics details and an estimate of the smallest device, see the reference below.

Bernstein, J. *"Heisenberg and the Critical Mass."* American Journal of Physics 70 (2002): 911–916.

## 89. Fabric of Space-time

Space-time is not a piece of cloth, nor does space-time behave like a real material. The metaphor "fabric of space-time" allows one to visually model space-time with its coordinates of space and time in a way similar to the simpler two-dimensional construction of woven cloth. There is no way for space-time to rip or tear, although mathematically there can be singularities of various dimensions and other mathematical properties that some people have stretched into real physical properties in the name of dramatic license.

As far as being able to go back in time as a time traveler, the time dimension of space-time is not like the spatial dimensions of space-time. Mathematically, the passage of time operator in quantum mechanics is antiunitary, while the spatial displacement operator is unitary. In addition, no one has shown convincingly that a fundamental particle can progress either forward or backward in time differently than we now experience and understand—that is, time passes at its normal rate for all particles and collections of particles. One possible explanation for all particles behaving the same with regard to the passage of time relies on the direction of time being built into the quantum state definition of a fundamental particle, with the opposite direction of time for its antiparticle.

The phrase "speed of gravity" is meaningless in the simplest interpretation, being an error in stating the *acceleration* of gravity. Most films confuse the concepts of speed and acceleration in the same manner that most people do. Unfortunately for society, the scriptwriters did not learn their beginning physics, and we all continue to suffer from their intuitive mistakes in describing the behavior of nature. Or perhaps a more positive view should be taken in that the entertainment value is improved by ignoring the laws of physics!

# Chapter 5
# Go Ask Alice

## 90. Spotlight

Yes, a spot of light from the lighthouse beacon when moving across your field of vision can move faster than c. But the actual light itself (i.e., the photons in the beam) moves from the source to the reflecting spot in the sky at c, no faster.

A good example from astrophysics is the radio wave beam from the pulsar in the Crab Nebula, which sweeps across our Earth observatory thirty times a second from a distance of a few thousand light-years. The electromagnetic waves coming to us from the distant source travel at light speed, but the sweep across our planet moves faster than c.

Occasionally one encounters other suggested examples of spots traveling faster than light speed, such as the intersection edge in a very long pair of scissors progressing outward when closing. Unfortunately, this intersection edge's speed is limited by the speed of sound in the metal of the scissors, which is quite slow compared to c. However, the spot of light on an oscilloscope trace can move across a screen faster than c even though this spot is produced by the slower-moving electrons striking a phosphor.

Bergmann, P. G. "Can a Spot of Light Move Faster than c?" The Physics Teacher 19 (1981): 127.

Rothman, M. A. "Things That Go Faster than Light." Scientific American 203, no. 1 (1960): 142–152.

———. "Not So Fast." Scientific American 269, no. 6 (1993): 10.

# 91. Quasar Velocity

The special theory of relativity says that *information* cannot be transmitted faster than light. The photons always travel at light speed in the local reference frame, but the space in which the photons travel may be expanding. An analogy would be a 100-meter race with the track lengthening during the race. The elapsed time to reach the 100-meter tape depends on the model for the expansion rate for the runner and for the photons from that distant quasar. Under these expansion conditions, there can be recessional velocities greater than c!

Chown, M. "All You Ever Wanted to Know about the Big Bang." New Scientist (April 17, 1993): 32–33.

Peebles, P. J. E., D. N. Schramm, E. L. Turner, and R. G. Kron. "The Evolution of the Universe." Scientific American 271, no. 4 (1994): 52–57.

———. "Out with the Bang." Scientific American 272, no. 3 (1995): 10.

Stuckey, W. M. "Can Galaxies Exist within Our Particle Horizon with Hubble Recessional Velocities Greater than c?" American Journal of Physics 60 (1992): 142–146.

# 92. Spaceship Approach

The observer sees the highly relativistic object approaching back side first! Therefore, the spaceship seems to be approaching tail first! What is often referred to as a contraction in the

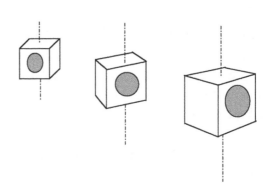

direction of motion for a relativistically approaching object is actually a rotation known as the Terrell effect.

We need to discuss some aspects of the Terrell effect to explain the behavior of the approaching spaceship. Consider a solid, opaque cube approaching. At low speeds, the light rays emitted off the back side of the approaching cube cannot pass through the box to reach the observer. At higher speeds nearing the speed of light, however, enough of the box moves out of the way for light emitted from part of the back side to reach the observer. When this behavior happens, the observer will not see all of the front side because some of the light rays from the front are intercepted by the extremely fast moving box. The box appears rotated, with the away side of the front hidden, and the near side of the back visible. The rotation angle increases with increasing speed, nearing c and with proximity to the trajectory. Additional complications also occur, such as nonrigidity, which we ignore in this simple explanation.

So the spaceship approaching at near light speed will appear rotated so that the back end is almost totally visible and the front end is almost totally hidden from view. J. Terrell in 1959 was the first to recognize that what physicists had been calculating as a Lorentz-Fitzgerald contraction is

actually a rotation for a real three-dimensional object. What we have described above is a snapshot of the spaceship (and cube)—that is, what photons from different parts of the object would imprint on a camera sensor simultaneously. E. Sheldon (see the reference below) discusses the stereoscopic appearance of a three-dimensional object that involves shearing and other distortions in addition to rotation, all these effects first discussed by J. Terrell.

Sheldon, E. "The Twists and Turns of the Terrell Effect." American Journal of Physics 56 (1988): 199–200.

Terrell, J. "Invisibility of the Lorentz Contraction." Bulletin of the American Physical Society 5 (1960): 272.

Weisskopf, V. F. "The Visual Appearance of Rapidly Moving Objects." Physics Today 13 (1960): 24–27.

## 93. Mass and Energy

The answer to both questions is equation 1, although the majority of physicists seem to prefer equations 2 or 3! Their choices probably are caused by the confusing terminology widely used in the physics literature that says that a body at rest has a "proper mass" or "rest mass" $m_0$, and a body in motion has a "relativistic mass" $m = m_0/\sqrt{(1 - v^2/c^2)}$.

There is only one mass in physics, m, which does not depend on the reference frame. This mass m is the relativistic invariant quantity in $E^2 - p^2c^2 = m^2c^4$, whereas the energy is different in different reference systems. There is no need to place the index 0 with the mass. However, the total energy $E$ needs the 0 index if the particle has no momentum in that reference frame—that is, $E_0 = mc^2$.

For a complete and stimulating discussion of these ideas and their history see the L. V. Okun reference below.

Einstein, A. "Zur Elektrodynamik bewegter Körper." Annalen der Physik 17 (1905): 891.

———. "Ist die Trägheit eines Körpers von seinem Energieinhalt abhängig?" Annalen der Physik 18 (1905): 639. Translated by W. Perrett and G. B. Jeffery in The Principle of Relativity. New York: Dover, 1923, p. 71.

———. "Zur Theorie der Brownschen Bewegung." Annalen der Physik 19 (1906): 371.

Okun, L. V. "The Concept of Mass." Physics Today 42, no. 6 (1989): 31–36.

# 94. Strain Gauge

The strain gauge continues to show a zero value. What I interpret as a length contraction when I run past is really the measurement of the length component along my direction of motion of the metal bar that appears to be rotated. The atoms do not move closer to one another, so the strain gauge remains at zero.

The apparent rotation is called the Terrell effect: if a snapshot is taken of a moving object, the object does not appear contracted, but rather rotated. A snapshot is understood to be a two-dimensional, nonstereoscopic photograph. The stereoscopic appearance of a three-dimensional object is more complicated because shearing and other distortions can be present. In fact, there is no such thing as a rigid object in relativity!

DeCampli, W. M. "A Gedanken Experiment to Demonstrate Lorentz Contraction." The Physics Teacher 13 (1975): 420–422.

Terrell, J. "Invisibility of a Lorentz Contraction." Bulletin of the American Physical Society 4 (1959): 294.

———. "Invisibility of a Lorentz Contraction." Physical Review 116 (1959): 1041–1045.

———. "The Terrell Effect." American Journal of Physics 57 (1989): 9–10.

# 95. Mass/Energy

Mass *is* energy. There is no distinction to be made. There is no conversion! In 1905, Einstein states explicitly: "The mass of a body is a measure of its energy content . . . ". What Einstein was stating is that mass and energy are *equivalent,* that they are possibly two different aspects of the same physical quantity, only their units have been chosen to be different. One does not *convert* one to the other if they are equivalent.

We can imagine a conversation with a student: "Does a photon have mass?" asks the student. "Yes, because the photon has energy." The student counters, "But for a photon $E = pc$, so the relation $E^2 - p^2c^2 = m^2c^4$ becomes $E^2 - p^2c^2 = 0$. Therefore m = 0 for the photon." Can you complete this dialogue?

Einstein, A. *"Ist die Trägheit eines Körpers von seinem Energieinhalt abhängig?"* Annalen der Physik 18 (1905): 639. Translated by W. Perrett and G. B. Jeffery, in The Principle of Relativity. *New York: Dover, 1923, p. 71.*

————. Relativity: The Special and General Theory. *New York: Crown, 1961, p. 47.*

# 96. System of Particles

No and yes! Except in the special circumstance described below, the answer is no. Energy and momentum are additive, but not mass. Mass is a measure of the magnitude of the energy-momentum 4-vector. From the total energy $E$ and the total momentum $P$ can be determined the mass $M$ of the system: $M^2c^4 = E^2 - P^2c^2$. Therefore the mass $M$ of the system is greater than the sum of the masses of its particles by the amount equal to the total kinetic energy of all the particles as seen in the frame in which the total momentum is zero. The exception "yes" occurs when all the particles move in the same direction with the same speed—that is, have the same velocity.

The value of defining the mass in this relativistic fashion means that $M$ determines the system's inertia, its resistance to acceleration by a force that acts on the system as a whole. A box with a hot gas of particles has more mass than the same box after the gas has cooled. Also, the box of hot gas exerts a greater gravitational pull on a test particle. In addition, a box of photons exerts a gravitational pull on a test particle, and vice versa.

Taylor, E. F., and J. A. Wheeler. Spacetime Physics: Introduction to Special Relativity. *San Francisco: W. H. Freeman, 1992, p. 135.*

# 97. Light Propagation

According to the special theory of relativity (STR), (1) no object can move at the speed of light, (2) the speed of light is the same for all observers, and (3) the space-time interval $\tau$ between two events defined by $\tau^2 = c^2\,\Delta t^2 - \Delta x^2 - \Delta y^2 - \Delta z^2$ is the same for all observers, but the $\Delta t$ and $\Delta x$ may be different, for example.

For one-dimensional motion $\tau^2 = c^2\,\Delta t^2 - \Delta x^2$. The driver has $\Delta x \neq 0$, so her $\Delta t$ must be greater for the two events than the elapsed time for the observer on the ground. Therefore the driver measures the longer time interval between events A and B.

Suppose the car makes a second run at a great speed. The nearer to the speed of light the car goes, the smaller is $\Delta t$ for the ground observer, and $\tau = c \Delta t$ is smaller in this case also. But again, as expected, the time interval is longer for the driver. The nearer to the speed of light the car goes in the ground frame, the difference will be the difference in arrival times as observed in the two frames.

*Taylor, E. F. "Light Propagation." The Physics Teacher 25 (1987): 252.*

# 98. Sagnac Effect

No, they do not tick at the same rates. Their tick rates are different because Earth is rotating with respect to an inertial reference frame such as the distant stars. The clock moving eastward has a higher velocity with respect to the inertial frame than the clock moving westward at all moments. According to the STR, the higher the velocity, the slower the clock ticks. That is, a clock ticks fastest when at rest in an STR inertial reference frame.

The difference in the elapsed time for the two clocks can be calculated by considering a light clock following a circular light path around the Equator.

One also could use a regular $n$-gon of flat mirrors to reflect the light around the Equator and then take the limit as $n$ becomes infinite. The light leaves from point P on the Equator of the rotating Earth and returns to point P in time T. The light going eastward has traveled the distance $2\pi R + \omega RT$ in the inertial system, where $\omega$ is the angular frequency of rotation with respect to the inertial reference frame. The point P has traveled $\omega RT$. The ratio of point speed to light speed is $\omega R/c = \omega RT/(2\pi R + \omega RT)$, from which $T = 2\pi R/(c - \omega R)$. For the system at rest, $T = 2\pi R/c$. Hence, when $\omega \neq 0$, define $\delta T = T - 2\pi R/c$ as the extra time required. Substitution for T gives $\delta T = 2\pi\omega R^2/[c\,(c - \omega R)]$. Upon returning to point P on the Equator after one circuit, the clocks will differ by $2\delta T$ for the measured elapsed times.

*Schlegel, R. "Comments on the Hafele-Keating Experiment." American Journal of Physics 42 (1974): 183–187.*

# 99. Light Flashes

The observer on planet A sees the flashes 20 minutes apart. From the STR postulate, we know that no observations of the light flashes only can discern which inertial frame is at rest. If the flashes sent out by the spaceship at 10-minute intervals are seen at planet B separated by 5 minutes, then flashes sent out from B at

10-minute intervals will be seen on the spaceship at 5-minute intervals.

One also realizes that if there were a light flash every 5 minutes from planet A, the observer on planet B would see them at 5-minute intervals. What is the interval for these flashes from A as seen by the spaceship observer? The answer is every 10 minutes, by invoking the STR postulate above. So the spaceship sees planet A's flashes to be spaced twice as much apart as the interval at the source on A; likewise, the 10-minute flash intervals from the spaceship must be twice the interval at planet A, or 20 minutes apart.

*Hewitt, P. G. Conceptual Physics, 6th ed. Glenview, Ill.: Scott, Foresman, 1989, pp. 650–656.*

# 100. Forces and Accelerations

No. In the STR, all contact forces will produce an acceleration in a direction *not* parallel to the applied force! For example, a rigid sphere is moving along the plus x-direction of an inertial reference frame. Now let an applied contact force act in the plus y-direction to increase the speed of the sphere in the y-direction. What happens to the speed in the x-direction? The x-component of the speed *decreases*—the object slows down in its original direction, corresponding to a negative acceleration!

To understand why the object slows in the x-direction when the applied contact force is in the y-direction, we begin with the space-time interval: $(\text{interval})^2 = c^2 \, \Delta t^2 - \Delta x^2 - \Delta y^2 - \Delta z^2$. For real objects traveling at speeds less than c, the time term is much larger than the spatial terms, and the interval is called the proper time $\tau$. The linear momentum $p^x$ in the x-direction in Newtonian physics is defined as $p^x = m \, dx/dt$ (for an object that is not changing its mass, i.e., excluding objects such as a leaking bucket of water). The correct STR expression simply substitutes proper time $\tau$ for Newtonian time $t$ so that $p^x = m \, dx/d\tau$. For a low-velocity object, $d\tau \sim dt$. But the actual relationship between $\tau$ and $t$ depends on the magnitude of the object's total velocity, a vector quantity, not just the speed component in the x-direction. Therefore, as the object speeds up in the y-direction, its speed in the x-direction must decrease to maintain a constant total velocity magnitude, otherwise its x-component of linear momentum would change, forbidden by the law of conservation of linear momentum. The pertinent relationship is $dt/d\tau = 1/\sqrt{(1 - v^2/c^2)}$. Remember, the mass is fixed in value.

One could write down the relativistic momentum $p^x = mv^x//\sqrt{(1 - v^2/c^2)}$ and argue that since m is constant, the component of velocity in the original direction must decrease to keep the momentum component constant.

Ficken, G. W. Jr. *"A Relativity Paradox: The Negative Acceleration Component."* American Journal of Physics *44 (1976): 1136–1137.*

González-Díaz, P. F. *"Relativistic Negative Acceleration Components."* American Journal of Physics *46 (1978): 932–934.*

Tolman, R. C. Philosophical Magazine *22 (1911): 458.*

# 101. Uniform Acceleration

In the STR, the velocity in the lab frame is no longer $V = a't$ for a uniform acceleration $a'$ in the moving frame. However, in the moving frame at each instant the expression $V' = a't'$ continues to be true. To convert from the moving frame to the lab frame, we must essentially convert the clock readings and time interval using $dt/d\tau = 1/\sqrt{1 - v^2/c^2}$. Here, $\tau$ is the proper time—that is, the clock reading on a wristwatch worn by an observer on the spaceship, say, and $d\tau$ is the proper time interval between two events at the *same* location. In the example, $\tau$ is the elapsed time on the wristwatch of the person on the moving frame. Hence, on the moving spaceship frame, $V' = a'\tau$.

Before we determine the answer for the velocity of the object in the lab frame, let's review the simpler problem of how velocities are added in relativistic frames. If an object moves forward with the velocity $V'$ in the spaceship frame, then the object's velocity $V$ in the lab frame is determined by the law of addition of velocities $V/c = (V'/c + V_s/c)/(1 + V'V_s/c^2)$, where $V_s$ is the uniform velocity of the spaceship in the lab frame. One can check the limiting case for low velocities, when $V'V_s/c^2$ is very small, to verify agreement with Galilean relativity—that is, the two velocities simply add.

To relate the acceleration of the object as seen by both observers, the addition of velocities expression is differentiated with respect to the time in the lab reference frame to obtain $a = a'/\{(1 + V'V_s/c^2)\sqrt{(1 - V_s^2/c^2)}\}^3\}$, a messy expression. The velocity of the accelerating object in the lab frame is found by substituting $V' = a'\tau$. Therefore $a \neq a'$ and $V < c$.

An alternative mathematical technique using a velocity parameter defined in terms of hyperbolic functions is given in the Taylor and Wheeler reference below.

Taylor, E. F., and J. A. Wheeler. Spacetime Physics: Introduction to Special Relativity. *San Francisco: W. H. Freeman, 1966, pp. 47–58.*

Tipler, P. A. Modern Physics. *New York: Worth, 1978, p. 27.*

# 102. Long Space Journey

The 7,000 light-year journey with 40-year aging is possible in STR physics but not in Newtonian physics!

Define $V/c = \tanh \theta$, where tanh is the hyperbolic tangent. Substitute into the law of addition of velocities to obtain $\tanh \theta = (\tanh \theta' + \tanh \theta_s)/(1 + \tanh \theta' \tanh \theta_s)$. Some checking of the mathematics of hyperbolic functions will reveal that the $\theta$s are additive, just as velocities are additive in Newtonian physics with Galilean relativity. That is, $\theta = \theta' + \theta_s$. Some people call $\theta$ the velocity parameter.

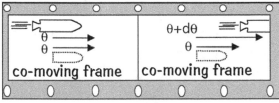

Astronaut time $\tau$     Astronaut time $\tau + d\tau$

Back to the problem at hand: How much velocity $V$ in the lab frame does the accelerating spaceship have after a given time? We need three frames of reference: the lab frame, the spaceship frame, and an instantaneously comoving inertial frame that for an instant has the same velocity as the spaceship. With respect to the instantaneously comoving frame, the velocity parameter changes from 0 to $d\theta$ in wristwatch time $d\tau$. In the *same astronaut time* the velocity parameter of the spaceship with respect to the lab frame changes from $\theta$ to $\theta + d\theta$. But $d\theta = a\, d\tau/c$. That is, each time interval $d\tau$ on the astronaut's wristwatch is accompanied by an additional increase $d\theta = a\, d\tau/c$ in the velocity parameter of the spaceship. Since the spaceship starts from rest, we get $\theta = a\tau/c$, telling us the velocity parameter $\theta$ of the spaceship in the lab frame at any time $\tau$ in the astronaut's frame.

Our solution is $V = c \tanh (a\tau/c)$. There is no limit to the product $a\tau$, which can be much greater than c, but $\tanh \leq 1$, so the lab velocity $V$ only approaches c after a long wristwatch elapsed time. The distance traveled in the lab frame is $dx = \tanh (a\tau/c)\, c\, dt$. In the lab frame, the astronaut's wristwatch seems to be ticking slower than the lab clock, so $dt = \cosh \theta\, d\tau$, with $\theta = a\tau/c$. Therefore $dx = c \sinh (a\tau/c)\, d\tau$, which can be integrated from zero astronaut time to the final time $T$ to produce the distance traveled $x = (\cosh (aT/c) - 1)\, c^2/a$.

The journey would be done by accelerating to the halfway distance at 3,500 light-years, then decelerating to the 7,000-light-year distance. Substituting 3,500 light-years in units of meters, g as 9.8 m $s^{-2}$, and the speed of light, one calculates a journey duration for the space traveler's wristwatch of $T \sim 8.62$ years. The round-trip would require about 34.5 years. So the space travelers would age less than 40 years!

Are there any plans to make this journey? Assuming human volunteers are available who want to achieve this feat, other factors, such as a reliable food supply, sufficient health care, and an energy source for the constant 1-g acceleration for 40 years would be difficult to provide with present technology. And, of course, more than 14,000 years would have passed for civilization here on Earth. Who or what would be here to greet them on their return?

*Taylor, E. F., and J. A. Wheeler.* Spacetime Physics: Introduction to Special Relativity. *San Francisco: W. H. Freeman, 1966, pp. 47–58.*

# 103. Head to Toe

Yes, your feet and toes age slower than your head. That is, whenever you are standing or sitting, a clock at the altitude of your head will tick faster than an identical clock at the altitude of your toes. The ambient gravitational field affects the tick rate of all clocks in the same way. A clock will tick fastest at rest in an inertial reference frame. The difference between clock rates in different gravitational environments is normally minuscule but measurable

and, to a first approximation, the time interval between ticks differs by $(\delta r/r)$ $GM/rc^2 \, \Delta T$, where $\delta r$ is the altitude difference, M is Earth's mass, $r$ is the radial distance from the center of Earth, G is the gravitational constant, c is the speed of light, and $\Delta T$ is the time interval between ticks on the reference clock. Substituting $r = 6.37 \times 10^6 \, m$ and $dr \sim 1.5 \, m$ produces a value of $1.6 \times 10^{-16} \, \Delta T$, an incredibly small change in rate. Over a lifetime of about 80 years, the head becomes about 0.4 microsecond older than the toes.

To understand the effect of gravitation on the clock rate, we can utilize the equivalence between an accelerating rocket frame and being in a uniform gravitational field. Consider two light flashes sent from the bottom of the accelerating rocket to its top, as shown in the animation diagram from the view of our inertial reference frame with respect to the stars. The two light flashes are one second apart in our frame but arrive at the top of the rocket three seconds apart. Why? Because the top has moved away from the approaching light flash with the appropriate acceleration value. Therefore the frequency of arrival is lower than the starting frequency. In a stroke of genius, Einstein realized that the only reason for different flash frequencies would be if the clock at the top ticked at a different rate than the

identical clock at the bottom. Therefore, gravitation makes time run slow.

Is there a place where one can put a clock so that the time interval between ticks becomes infinite? Yes, near a black hole, at the event horizon.

Hewitt, P. G. Conceptual Physics, *6th ed.* Glenview, Ill.: Scott, Foresman, 1989, pp. 671–678.

Taylor, E. F., and J. A. Wheeler. Spacetime Physics: Introduction to Special Relativity. San Francisco: W. H. Freeman, 1966, p. 154.

# 104. Neutrino Mass

For a change in a system to occur—such as the change of a muon neutrino to an electron neutrino, for example—time must elapse. That is, the reference clock must tick in the rest frame of the muon neutrino. We know that the greater the velocity of a real clock in our laboratory reference system, the slower is its ticking rate. In the speed limit of a massless particle such as a photon traveling at light speed, the clock would not tick. As a photon traverses the universe, no time elapses in its reference system. The photon can be absorbed by an atom and disappear, but the photon cannot change directly into another photon. Likewise, if all three neutrino types did not have any mass, none could oscillate into another neutrino type because they do not experience the passage of time. Therefore, for neutrino oscillations to occur, at least two neutrino types must have mass. The data indicate that the sum of the three neutrino masses cannot exceed about 1 $eV/c^2$, very much smaller than the 0.511 $MeV/c^2$ mass of an electron.

# 105. Spaceship Collision

The method of determining position and clock reading for the three events first before answering the question is a good one. However, the values inserted already are not all correct for the observer. Simultaneous measurements at both the origin $X_1 = 0$ and at $X_2 = L$ cannot be made by the method assumed since they are not equidistant. Therefore, if the notation $(X, T)$ is correctly $(0, 0)$ for event 1, then event 2 is labeled by $(L, -L/c)$ because the light from event 2 takes $L/c$ seconds to travel the distance $L$ to the observer. Event 3 is *not* at position $L/2$ between the two spaceships at $T = 0$ because spaceship B has already traveled for $L/c$ seconds. Therefore the distance between the two spaceships is $L - VL/c$. Thus $T_3 = L(1 - V/c)/2V$. We can summarize the events as:

Event 1:   $X_1 = 0$                     $T_1 = 0$

Event 2:   $X_2 = L$                     $T_2 = -L/c$

Event 3:   $X_3 = L(1 - V/c)/2$   $T_3 = L(1 - V/c)/2V$

These same events can be specified in the inertial frame (primed) of spaceship A as:

Event 1′:   $X_1′ = 0$                          $T_1′ = 0$

Event 2′:   $X_2′ = \gamma L(1 + V/c)$       $T_2′ = -\gamma L(1 + V/c)/c$

Event 3′:   $X_3′ = 0$                          $T_3′ = \gamma^{-1} L(1 - V/c)/2V$

We have defined $\gamma = \sqrt{(1 - V^2/c^2)}$ and have used the normal Lorentz transformations $x′ = \gamma (x - Vt)$ and $t′ = \gamma (t - Vx/c^2)$ of the STR.

Now, finally, we can determine the clock reading—that is, the elapsed time—for the observer who sees the collision a distance $L(1 - V/c)/2$ away as $T = L(1 - V/c)/2V + L(1 - V/c)/2$, which reduces to $T = L(1 - V^2/c^2)/2V$. The observer on spaceship A has an elapsed time of $\gamma^{-1} L(1 - V/c)/2V$.

Chai, A.-T. "Some Pitfalls in Special Relativity." American Journal of Physics 41 (1973): 192–195.

## 106. Twin Paradox

Peter experiences actual accelerations during his spaceship journey that will result in less aging than for his twin brother, who has remained at home on Earth. Even if the acceleration was simply an immediate turnaround at the farthest distance, the spaceship velocity vector reversed direction from $+V$ to $-V$, a change of $2V$, in a time interval $T$. Peter felt the acceleration. Therefore, all observers will agree that Peter was the traveler and that his clocks ran slow, so he ages less than his stay-at-home twin.

Feynman, R. P., R. B. Leighton, and M. Sands. The Feynman Lectures on Physics. Vol. I. Reading, Mass.: Addison-Wesley, 1966, pp. 16-3 to 16-4.

# Chapter 6
# Start Me Up

## 107. Air-Driven Automobile Engine

Yes. Many companies worldwide have been operating compressed-air-driven cars using a standard gasoline four-cylinder engine but replacing the gasoline fuel input with compressed air from a tank. Of course, there is no combustion, so the electrical supply for the spark plugs is not needed, nor will there be any need to change the oil very often. The compressed-air tank is stored in the trunk.

The piston upstroke compresses

and heats the atmospheric air in the cylinder chamber until just about top dead center, when cool compressed air is injected to drive the piston down and turn the crankshaft. The process repeats itself until the compressed air is depleted. The exhaust is just cool air. The horsepower rating is about 35 horsepower for some models, but the value will increase to more practical values with further development. Using traditional electricity sources to compress the air, there will be some carbon dioxide air pollution for the overall process, but only about a fifth or less that of conventional autos.

Perhaps the best-known air-powered car is that designed by French inventor and engineer Guy Nègre for Motor Development International (MDI) in France. The car has a maximum speed of about 110 kilometers per hour and can travel about 300 kilometers at a cost of less than a cent per kilometer. (Details can be found on the Internet.)

## 108. Coin Tosses

You should be able to pick out the experimentally obtained sequences with about 98 percent accuracy! In a random sequence of 256 fair coin tosses, you would expect to find at least 1 run of 6 heads or 6 tails with a probability of 98.2 percent. If the sequences imagined by students unfamiliar with

the characteristics of randomness do not contain long runs, you should be able to distinguish them reliably.

The actual estimate of the number of runs with 6 or more heads or tails is 4, meaning that you should be able to find about 4 of these long runs. For a run of at least $k$ heads in $n$ tosses, where $k \geq 1$, the mean number of runs is $\sim n/2^{(k+1)}$; thus $2 \, (256/2^7) = 4$. The following table contains actual data for 256 coin tosses, with a 1 representing heads. You can count the numbers of the different run lengths.

```
1011111010110110101101001000110 0
1101111010000100100111001010010 0
1100110011111000100000101111100 0
1011001000111110011011100111001 0
1111100001101110000000101111100 0
1111011011000000101000001011111 0
1111110011101100101110001011111 0
0111011011110000111111000001100
```

Silverman, M. P. A Universe of Atoms, an Atom in the Universe. *New York: Springer-Verlag, 2002, pp. 284–291.*

## 109. More Coin Tosses

Most people would expect to return to the lamppost quite often—20 or more times during the 1,000 tosses. However, returning more than 2 times is unlikely! There will be a long drift away from the lamppost for most of the coin-tossing time.

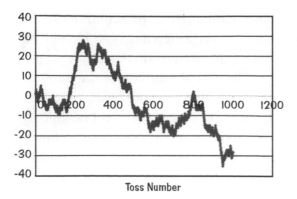

Toss Number

One can do the actual coin tossing to experience the drift away from the origin for long time periods, or one can run a computer simulation. The expected distance after $N$ tosses will be $\sqrt{N}$ times the unit step distance, the random walk distance in one dimension.

## 110. Brownian Motor

As long as the ratchet potential is off, there can be no net movement to the right or the left because the particles will move diffusively according to a (biased) random walk, leading to a variance in position of $\delta x = \sqrt{(2D\tau)}$ and a mean position of $<x> = f\tau/\gamma$, where $D = kT/\gamma$ is the diffusion constant. When the ratchet potential is switched on, one or more particles get trapped in one of the potential minima. If $\alpha L \geq \delta x \geq (1 - \alpha)L$ for the variance holds, the particle on average gets trapped into the minimum left of the starting point. The maximum flux is obtained if the switching time t is large enough to assure that the particle can adjust in the trapping minimum (adiabatic adjustment time) and also is small enough to fulfill the above requirement for the variance. Roughly, one can say that a net flux to the left always occurs when thermal energy is significantly smaller than the potential maximum, the external force chosen is not too big, and the driving frequency matches the adiabatic adjustment time needed for the particle to move into a potential minimum.

Where does the energy come from leading to a drift against the external force? The energy does not come from the heat bath but from the ratchet potential when it is switched on. At that moment the potential energy of the particle will suddenly be increased. In a simulation, this can be seen by a sudden increase of the energy. But most of the energy pushed into the system will just be dissipated into the heat bath due to the relaxation of the particle into a potential minimum. Only a tiny portion will be used for doing work. Thus a Brownian motor does not violate any law of thermodynamics because it only turns one type of work into another one. Nevertheless, the fluctuating force due to the heat bath is essential for a Brownian motor.

For more details and possible applications in biology and chemistry read the following review articles. For a simulation, there are Java applets on the Internet.

Astumian, R. D. "Thermodynamics and Kinetics of a Brownian Motor." Science 276 (1997): 917–922.

———. "Making Molecules into Motors." Scientific American 285, no. 1 (2001): 57–64.

# III. Magnetocaloric Engine

The ferrofluid is cycled around the loop by the stationary permanent magnet. A small volume of ferromagnetic material has less energy where the magnetic field density is greater, just like iron filings are pulled to the poles of a magnet. So the ferrofluid approaching the magnet becomes magnetized and drawn into the loop volume between the magnetic poles. But the heat source nearby warms the ferrofluid to partially randomize the magnetic dipoles in the ferrofluid, so the energy of the system can be lowered again by drawing in some more of the cooler magnetized ferrofluid, which pushes out the warmed ferrofluid. The heat put in by the heat source is deposited at the heat sink, and the cycle repeats.

To use this engine for the solar heating of buildings, two avenues of operation are possible. One could have all the piping contain ferrofluid, which probably would be costly. The alternative would be to have a small closed loop of ferrofluid in contact with a large loop of piping containing the water to be heated in a heat exchange device. The advantage over typical systems would be no moving mechanical parts in the solar heating system.

Rosensweig, R. E. "Magnetic Fluids." Scientific American 247, no. 4 (1982): 136–145.

# 112. Magnetorheological Fluid

The flow properties of the fluid change so radically that the fluid becomes gel-like and can be pushed to one side of the beaker where no relaxation may occur. The degree of solidification depends on the inherent properties of the fluid and the strength of the magnetic field. Of course, its solidification may vary within the gel itself because the magnetic field may vary with position in the beaker. Practical applications of these materials with their unusual properties are being devised and tested. Perhaps automobile braking systems may someday use these types of fluids to replace solid materials that wear away.

Klingenberg, D. J. "Making Fluids into Solids with Magnets." Scientific American 269, no. 4 (1993): 112–113.

## 113. Binary Fluids

Both phase diagrams can represent actual binary fluids, although the diagram to the right is quite rare. To understand these phase diagrams, both energy and entropy must be considered. The energy part involves the van der Waals interaction between adjacent molecules, an induced dipole-dipole electromagnetic interaction. In general, this attractive force between unlike molecules is much weaker than the attractive force between like molecules. The stronger the force of attraction holding the molecules together, the lower the energy of the system. Hence, when most of the molecule's neighbors are of the same chemical species, the system energy is lowest and immiscibility is favored. Even the increased random tumbling about of the molecules at higher temperatures doesn't disrupt this clustering of like molecules.

However, energy considerations alone do not explain the behavior of binary liquids. Why are they miscible at all? The miscibility occurs at lower temperatures because the system tends to minimize not its energy but rather its free energy, $E_{free} = E_{sys} - TS$. The free energy is the energy of the system minus the product of the temperature $T$ and the system's entropy $S$. At a given $T$, the free energy can be decreased by decreasing the system's energy or by increasing its entropy. At low $T$, changing the entropy has a minimal effect because the product $TS$ may be small. But at high $T$, the product can be large. So systems at high $T$ tend to maximize their entropy, that is, their randomness or disorder.

We now have a good argument for ruling out the diagram to the right, with its reappearing miscible phase at low temperatures. Not so! For some molecules, hydrogen bonding occurs with its very small angular spread, locking two molecules together. This hydrogen bonding occurs primarily at lower temperatures because of the orientation dependence, with "orientation" entropy lost in forming the hydrogen bond being greater than the "compositional" entropy gained. Therefore both energy and entropy are lowered, and the lowered energy of the hydrogen bond has a large effect on the free energy. Water and butyl alcohol is one example of a binary liquid with the rare phase diagram.

Walker, J. S., and C. A. Vause. "Lattice Theory of Binary Fluid Mixtures: Phase Diagrams with Upper and Lower Critical Solution Points from a Renormalization-Group Calculation." Journal of Chemical Physics 15 (1983): 2660–2676.

———. "Reappearing Phases." Scientific American 256, no. 5 (1987): 98–106.

## 114. Baseball Bats

The main source of drag on the swing of a baseball bat is not air friction but the retarding force produced by the pressure difference across the bat from front to back. As the bat carves its swath, the air in front gets separated into two boundary layers that pass around the bat and recombine behind the bat. In the wake of the bat, between the two separated boundary layers, the "lack of air" means a lower pressure immediately in back of the bat, with a resulting backward force due to the pressure difference. Therefore, some of the energy of the swing does work against this backward force.

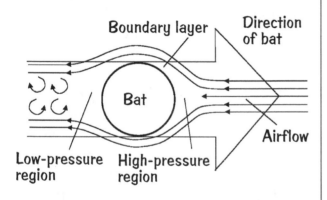

A dimpled bat sends the boundary layers tumbling in turbulent eddies into the space behind the bat, reducing the pressure difference and cutting the drag. More swing energy is now available to accelerate the bat and to transfer to the ball, so the ball's exit velocity will be increased.

*Gibbs, W. W. "To Fenway, with Love."* Scientific American *271, no. 1 (1994): 98.*

## 115. Old Glass

Many people have suggested that the glass experiences some flow downward in response to the gravitational pull of Earth. Contrary to popular conjecture, there is no evidence that any of this old glass could flow enough during the time interval of centuries to create the difference from top to bottom.

Another factor against the flow hypothesis is the actual profile, which is essentially a linear relationship of thickness to vertical distance. As a simple model, assume that the properties of the glass are identical at each vertical position along the pane. If a fixed amount of glass material flows from position 10, say, the same amount would replace this amount from position 11, slightly higher up the glass. The major changes over a long time interval would be a thick buildup at the very bottom and a depletion at the very top, with practically no thickness change between, in contrast to the linear dependence of glass thickness to height.

In the old days, window glass production made panes that varied slightly in thickness from one end to the other because the flat support surface had a slight tilt. The installers simply put the thicker end on the bottom. Quality control must have been marginal in some areas of the world, because we have seen some large differences in glass thickness between the two ends!

Glass is normally elastic at temperatures below about 1000 K, and glass may break but never deform permanently because the solid is crystalline. Delicate telescope and camera lenses would reveal such creep by changing their optical characteristics in obvious ways.

*Pasachoff, J. M. "Comment on 'Magnetic Fluids.'"* American Journal of Physics 66 (1998): 1021.

*Zanotto, E. D. "Do Cathedral Glasses Flow?"* American Journal of Physics 66 (1998): 392–395.

# 116. Ferromagnetism

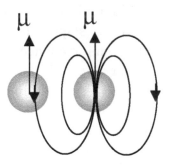

Many atoms and molecules have an inherent magnetic dipole moment. When we assume that each dipole behaves independently of its neighbors except for its alignment, the magnetic field direction next to the dipole is *opposite* to the direction in which the dipole itself points. In paramagnetic substances the dipoles are far enough apart to behave approximately independently, and when no applied field is present, these dipoles have random orientations. Each dipole is affected by the applied magnetic field but not by its neighbors. The applied magnetic field competes with the random thermal motion to cause a net magnetization that increases nearly linearly with the strength of the applied field, the ratio being known as the magnetic susceptibility.

When the density of magnetic dipoles becomes high enough for neighbors to affect each other, only neighbors in the head-to-tail configuration will tend to align one another. The side-by-side neighbors will be oppositely aligned because all the fields from its neighbors are opposite at its location. Thus every other dipole in each layer of the crystal will be aligned and form one sublattice, like the white squares on a checkerboard, and the remaining dipoles (on the black squares) will form a second sublattice of dipoles pointing in the opposite direction. The two sublattices interact strongly or ferromagnetically, but they cancel each other's magnetization. Therefore, when magnetic dipole moments are crowded together, they are more likely to disalign their nearest neighbors than to align them. So ferromagnetism is rare.

Then how can ferromagnetic substances exist at all? Via a *cooperative* effect when the dipoles are very close and no longer behave independently. In these conditions a state of lower energy can form if groups of dipoles align each other into magnetic domains that themselves point in random directions. With an applied field, these domains will change their sizes to find the lowest energy state. Of course, domain formation cannot form above a certain temperature called the Curie temperature, because the thermal agitation interferes with the dipole interactions. Above the Curie temperature the substance becomes paramagnetic.

*Kolm, H. H. "Why Are So Few Substances Ferromagnetic?" The Physics Teacher 20 (1982): 183–185.*

# 117. Coupled Flywheels

The overall angular momentum of the system must be conserved, so including just the change in angular momenta of the flywheels leads to an incomplete calculation. The tension is different in the two sides of the belt, so the belt exerts a downward force on pulley 2 and an upward force on pulley 1. These forces are counteracted by reactions at the bearings, in addition to the reactions to the weight of the components. These additional reactions produce a torque that accounts for the change in angular momentum.

If the pulleys are the same size, this additional torque does not exist unless the belts are crossed.

# 118. Superconductor Suspension

The demonstrated superconductor suspension does not illustrate the Meissner effect. Instead, this demonstration depends on the persistent eddy current in the zero resistivity superconducting material induced by the magnet. The eddy current direction is determined by Lenz's law to produce a magnetic field that ends up causing a repulsion between the superconductor and the permanent magnet.

To show the Meissner effect, the sequence of events must be different. Place the superconductor on the magnet at room temperature first, and then cool the superconductor below its critical temperature $T_c$. Then the magnetic flux will be "expelled" by the Meissner effect and the superconductor will become suspended above the magnet.

*Wake, M. "Floating Magnet Demonstration." The Physics Teacher 28 (1990): 395–397.*

# 119. Nanophase Copper

With smaller grain sizes, one would expect there to be many more grain boundaries in the nanophase copper metal than in normal copper. The extra grain boundaries would stop or impede any moving dislocation, thereby making the nanophase copper much harder. However, the surprise turned out to be that nanophase copper is mostly dislocation-free! Lacking large numbers of moving dislocations, these nanophase metals are much stronger.

Siegel, R. W. "Creating Nanophase Materials." Scientific American 275, no. 6 (1996): 74–79.

## 120. Head of a Pin

Experiments show surprising results. Any fraction of the fundamental charge, such as +0.5 e or –0.1 e, can exist on the head of a pin! The conceptual argument goes as follows. The metal pin is an electrical conductor. In general, an electric current flows in the conductor because some free electrons move through the lattice of atomic nuclei. Any particular volume of the conductor has virtually no charge because the negative charges are balanced by the positive charges of the nuclei.

So the important physical quantity is not the electric charge in any given volume but instead how much charge has been carried through the conductor—that is, the "transferred charge," which can have any value, even a fraction of the charge of a single electron. This "transferred charge" is proportional to the sum of the shifts of all the electrons with respect to the lattice of nuclei. These electrons in the conductor can be shifted as little or as much as desired, so the sum can change continuously, and therefore so can the "transferred charge." The pinhead can have any charge value, not just integer multiples of the fundamental charge.

Likharev, K. K., and T. Claeson. "Single Electronics." Scientific American 266, no. 6 (1992): 80–85.

## 121. Coulomb Blockade

No, the current across the junction will not be a steady current. There will be single electron tunneling (SET), with the voltage across the junction changing periodically with a frequency equal to the current divided by the fundamental unit of charge e.

The tunnel junction is a conductor-insulator-conductor device, so transferred charge flows through the conductor to accumulate on the surface of the electrode against the insulating layer of the junction. An opposite surface charge of equal amount accumulates on the other electrode across the junction. The actual amount of surface charge has a continuous change in value as the charge accumulates, including fractional values such as +0.8642 e, because the electrons near this surface can adjust their positions slightly.

However, only discrete amounts of charge can tunnel through the insulating layer—that is, each electron tunneling through changes the surface charge by +e or −e, depending on the direction of tunneling. The tunneling process is energy-dependent. If the charge at the junction is greater than +e/2, an electron can tunnel through to reduce the surface charge by e, thus reducing the electrostatic energy of the system. And if the surface charge is less than −e/2, an electron can tunnel in the opposite direction to decrease the energy. But if the surface charge value is greater than −e/2 or less than +e/2, tunneling would not occur because the system energy would increase. This tunneling suppression is known as the Coulomb blockade, first studied in the 1950s.

The tunnel junction connected to a constant current source begins in the Coulomb blockade condition, then reaches tunneling for the one-electron condition, then back to the Coulomb blockade, then one-electron tunneling, etc. The analogue may be a dripping faucet.

Many electronic devices are being made with SET operation. For example, an SET transistor can switch on or off the flow of billions of electrons per second when the charge on the middle electrode is changed by only half the charge of an electron!

*Likharev, K. K., and T. Claeson. "Single Electronics."* Scientific American 266, no. 6 (1992): 80–85.

# 122. Deterministic Competition

The time evolution here depends on the value of $r$. One finds that $N_t = 1$ is a stable equilibrium only when $r$ lies between 0 and 2. If $r = 2.3$ with $N_0 = 0.5$, then successive $N_t$ will oscillate between about 1.59 and about 0.40 as a stable 2 cycle. For $r > 3.102$, no cycle is stable, all cycles are possible, etc.

In the chaotic regimes, the equation results are deterministic, but the time evolution is indistinguishable from that governed by probability laws. One really needs to see the calculations proceed to appreciate the amazing behavior of this simple-looking equation.

*Gleick, J.* Chaos: Making a New Science. *New York: Penguin, 1987, pp. 166–186.*

# 123. Two Identical Chaotic Systems

Yes, the two identical chaotic systems described can be synchronized. Chaotic systems are very useful for several reasons: (1) Chaotic systems are a collection of many regular, ordinary behaviors, none of which dominate.

(2) The proper perturbation can encourage the chaotic system to follow one of its many regular behaviors. (3) Chaotic systems are very flexible because they can rapidly switch among different behaviors. (4) Chaotic systems are deterministic and, although no one can say which output will result, two identical chaotic systems of the appropriate type will produce the same output in response to the same signal input.

To synchronize two identical chaotic systems each with the stable subpart behavior, one can apply the appropriate pseudoperiodic signal (one type is called a Rössler signal) to coax them into step. For the reasons listed above, the outputs will be the same. The details can be learned in the reference below, where the chaotic attractor and the Poincaré section are discussed. Applications to secure communications and to biological systems are included also.

Ditto, W. L., and L. M. Pecora. "Mastering Chaos." Scientific American 269, no. 2 (1993): 78–84.

# 124. Tilley's Circuit

The galvanometer does nothing! There is no induced potential because no work was done (assuming frictionless switches). This result appears to violate Faraday's law $V = d\Phi/dt$, where $V$ is the potential difference induced by the rate of change of magnetic flux $\Phi$. But work must be done for $V$ to be generated because the change in the work $d\text{Work} = V\,dt$.

Nussbaum, A. "Faraday's Law Paradoxes." Physics Education 7 (May 1972): 231–232.

# 125. Thermal Energy Flow

The classical flow of thermal energy toward the cooler region occurs because the free energy of the combined system $E_{\text{free}} = E_{\text{sys}} - TS$ becomes less, where $T$ is the temperature and $S$ is the entropy. If the free energy is the same at two temperatures, one can see that for a given amount of system energy there is more disorder at the lower temperature. Assuming that the two-block system initially simply transfers thermal energy from the warmer to the cooler block, with no other energy transfers, then a cooler system is preferred.

Dyson, F. J. "What Is Heat?" Scientific American 191, no. 3 (1954): 58–64.

# 126. Cadmium Selenide

The wavelength of visible light is comparable to nanophase cluster sizes. For example, greenish light has a wavelength of about 580 nanometers, five to ten times the nanophase cluster sizes. Clusters behaving as particles ranging from about 1 nanometer to 50 nanometers in diameter are too small to have any significant scattering of visible light, so these materials are effectively transparent. Clusters of sizes comparable to particular wavelength ranges of visible light are subject to quantum confinement restrictions.

Quantum mechanics predicts the correct behavior at the small cluster sizes. The smaller the nanophase cluster size becomes, the greater are the energy spacings for the electron states. Which colors of light are absorbed and emitted are determined by these energy spacings. If the energy spacings are too great, the incoming light will not be absorbed, and light of that wavelength and longer will not be scattered. For example, a typical semiconductor is cadmium selenide. When the size of the cluster is 1.5 nanometers, the cadmium selenide appears yellow, but when the size is 4 nanometers, it will appear red. And larger clusters appear black. Therefore, the observed color of the clusters in the nanophase depends on their actual sizes.

## 127. Optical Solitons

Under the right conditions, the two effects—dispersion and the Kerr effect—can be made to cancel exactly. The nonlinearity of the Kerr effect can delay the "fast" carriers relative to the "slow" carriers, bringing them together to counter the dispersion.

These pulses conserving their shape and integrity are exhibiting soliton behavior. Optical solitons were first observed in fibers in 1980 and are now fundamental components in optical transmission systems.

Desurvire, E. "The Golden Age of Optical Fiber Amplifiers." Physics Today 47 (1994): 20–27.

## 128. Ceramic Light Response

Certain ceramic materials will change their shape upon exposure to light because some molecules in the material have changed their shape upon absorption of particular frequencies of light. If the responses of many molecules are coordinated, the overall effect can be a macroscopic shape change. Called the photostrictive effect, research began in the 1990s, and some practical devices are beginning to be developed, such as direct conversion of light to mechanical displacement for speakers instead of conversion to an electrical signal first. A telephone speaker could be one of the first products.

These ceramics are examples of a new type of "smart" material. The four most widely used classes of smart materials are piezoelectrics, electrostrictors, magnetostrictors, and shape-memory alloys. The resulting changes in the shapes of these materials are large enough to make them useful as actuators. A sensor receives a stimulus and responds with a signal; an actuator produces a useful motion or action. By definition, smart materials are both sensors and actuators, because they perform both functions.

Photostrictive materials such as PLZT—a combination of lead, lanthanum, zirconium, and titanium—someday may be used to control robots and machines. Engineers at Pennsylvania State University, for example, are exploring applications for devices that move when light shines on them and have created a two-legged stand that walks very slowly when illuminated.

Dogan, A., et al. "Photostriction of Sol–Gel Processed PLZT Ceramics." Journal of Electroceramics. 1, no. 1 (1997): 105–111.

Newnham, R. E., and A. Amin. "Smart Systems: Microphones, Fish Farming, and Beyond." ChemTech 29, no. 12 (1999): 38–46.

# 129. Random Movements

Wobbles in any system can be followed with fast cameras. For most human actions, from balancing a stick vertically on a finger to balancing on a tightrope, wobbles occur that last from seconds to tens of milliseconds. Usually the shorter the fluctuation, the more of them there are. But the typical human reaction time for such balancing acts is about 100 milliseconds, so most of the wobbles are faster than humans can react. Mathematical modeling of human balancing acts match the measured fluctuations only when the person or object is on the verge of falling. Then the random fluctuations cancel each other out and the object remains upright.

Related research has found that elderly people and others with balance problems showed signs of better balance when they stood on a pair of battery-operated, randomly vibrating insoles. The idea is that these vibrations amplify balance-related signals from the feet to the brain and vice versa that may have become reduced by age or illness. When people walk, then turn or reach, they are most vulnerable to a fall. When a person leans or sways to one side, the pressure on the sole of that side increases, and the nervous system senses the change in pressure and sends a message to the brain so that the posture can be adjusted. In many people, those messages can be altered by age, stroke, or conditions such as diabetes. Further testing is under way to optimize these helpful insoles.

Cabrera, J. L., and J. G. Milton. "On-Off Intermittency in a Human Balancing Task." Physical Review Letters 89 (2002): 158702.

Chow, C. C., and J. J. Collins. "Pinned Polymer Model of Posture Control." Physical Review E 52 (1995): 907–912.

Priplata, A. A., et al. "Vibrating Insoles and Balance Control in Elderly People." Lancet 362, no. 9390 (2003): 1123–1124.

# 130. Gravitational Twins

The traveling twin actually returns much younger than her stay-at-home sister. The argument given was correctly stated but incomplete. The local gravitational tidal effects are not the same for the twins—that is, the rate of change of gravitational potential experienced was different. These tidal effects contribute to the clock rates and, when included in the calculations, contribute enough to change the result so that the stay-at-home twin ages faster and is older upon return of her sister. For a calculation, see the reference below.

Bradley, M., and J. Higbie. Physics Teacher 22, no. 1 (1984) 34–35.

# 131. Photon Engine

We can analyze the operation of the quantum Carnot engine in the same manner in which we would analyze a classical Carnot engine. Let $Q_{in}$ be the energy absorbed from the bath atoms during the isothermal expansion and $Q_{out}$ be the energy given to the heat sink during the isothermal compression. Then the Carnot engine efficiency $\eta = (Q_{in} - Q_{out})/Q_{in}$.

If the bath atoms are assumed to be two-state systems that absorb and emit radiation at the same photon frequency, then we need the thermodynamic properties of a photon gas in order to determine the theoretical efficiency of this photon engine. Assuming thermal equilibrium for the photon gas, the average number of photons $n_2$ with energy $\varepsilon$ coming in from the heat bath at temperature $T_2$ is given by $n_2 = 1/(\exp[\varepsilon/kT_2] - 1)$, while the average number of photons $n_1$ leaving at temperature $T_1$ is $n_1 = 1/(\exp[\varepsilon/kT_1] - 1)$. Since $Q_{in} \propto n_2 \varepsilon$ and $Q_{out} \propto n_1 \varepsilon$, the efficiency of the quantum Carnot engine is $\eta = 1 - T_1/T_2$, exactly the same as for the classical Carnot engine. When there is only one heat bath, with $T_1 = T_2$, no work can be done.

A different quantum engine occurs when the bath atoms have three states instead of two, bringing in quantum behavior called quantum coherence, with a nonvanishing phase difference between the two lowest atomic states induced by a microwave field. One can eliminate the photon absorption process (analogous to laser operation without a population inversion). The temperature $T_2$ becomes altered to a different effective temperature, $T_\phi$. The efficiency $\eta_\phi = (T_\phi - T_1)/T_1$ can exceed the efficiency of the classical Carnot engine. This quantum engine can extract work from a single heat bath, even when $T_1 = T_2$! For the details of the three-state quantum engine's operation, see the reference below.

*Scully, M. O., et al. "Extracting Work from a Single Heat Bath via Vanishing Quantum Coherence." Science 299 (2003): 862–864.*

# Chapter 7
# A Whole New World

## 132. Grain of Sand

If one assumes that the grain of sand has a diameter that is a reasonable fraction of 1 millimeter, then the line of atoms would be about $10^{10}$ meters long, about thirty times the distance to the Moon!

## 133. Forensics

Until the mass production of paints became available in the late 1800s and early 1900s, each paint used by an artist is known to contain atoms in particular characteristic amounts, depending on the source. Paints were originally made from natural materials, so when an artist mixed his or her paints, there was usually a unique mixture of atoms and molecules for each color and color combination.

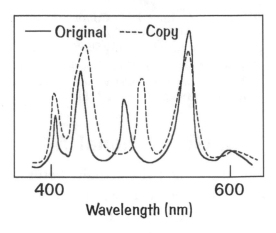

Different atoms absorb and emit their unique characteristic frequencies of light in the visible and the ultraviolet. The types of atoms present and the intensity of the characteristic spectrum from each atom type will create a "spectral fingerprint" for each artist. As you know, some artists simply laid out the design of the painting, for example, and lesser painters filled in the regions, with the master artist completing the final touches. Even these paintings have their own fingerprint of spectral colors.

With a tunable laser capable of scanning from the infrared frequencies to the ultraviolet frequencies, the "spectral fingerprint" of any region of the painting can be recorded and compared to other paintings by the same artist or even other artists, including fraudulent painters. This laser approach is normally combined with other approaches to achieve the comprehensive evaluation.

The laser technique also permits the identification and removal of environmental coatings on top of the paint beneath, such as dust and grime, and ensures that no harm to the painting occurs. Famous paintings such as Rembrandt's 1642 *De Nachtwacht* (The Nightwatch) in Amsterdam's Rijksmuseum have had the soot and grime safely cleaned off to reveal a marvelously brighter background of faces when compared to the somewhat obscure dull background that had existed for centuries.

# 134. Doppler Elimination?

Yes. First consider the emission process. Normally, a typical electric dipole emission occurs with a single photon exiting the atom as a result of an allowed transition within the atom that conserves energy and angular momentum—that is, the angular momentum of the atom changes by ±1 unit of Planck's constant $h/2\pi$. The probability for all other emission processes is lower by a factor of 1/137, or by a higher power of this factor.

A two-photon electric quadrupole emission process is possible between two atomic states with angular momentum quantum numbers differing by zero or two units of $h/2\pi$. There is a broad continuous spectrum of possible energies for the two photons emitted in this quadrupole emission process. A very small fraction of these two-photon emissions will spit out two photons of the same energy, go off in opposite directions, and produce no recoil of the atom. The two-photon emission from hydrogen was the first atom to be measured and the first to be calculated by quantum electrodynamics (QED) in the 1940s. Two-photon emissions after laser excitations have become commonplace for many uses in today's optics research.

Likewise, simultaneous two-photon absorption is possible. A container of single atoms is placed between two counterpropagating laser sources, shining two identical frequency laser beams on an atom so that energy and angular momentum will be conserved and recoilless absorption can occur. First achieved in the 1970s, the precise energy-spacing values within atoms have been determined. Today, two-photon absorption with nonidentical energies plays a critical role in the upconversion of laser light to higher frequencies to achieve coherent beams in the UV and for providing light sources of precise frequencies.

At the nuclear level, recoilless gamma-ray emission and absorption are possible if the whole crystal recoils simultaneously with the photon emission or absorption. This Mössbauer

Effect transition, discovered in the 1950s, relies on the inability in principle of identifying the single nucleus involved and includes an exponential factor proportional to the negative ratio of the temperature of the crystal to its Debye temperature.

As an interesting historical note, Albert Einstein in 1917 was among the first to recognize that classical electromagnetism cannot explain spontaneous emission of light from atoms. In particular, he inferred that an atom must recoil upon spontaneous emission, in conflict with the symmetric-field distributions produced by electromagnetic theory based on Maxwell's equations. According to Einstein, ". . . outgoing radiation in the form of spherical waves does not exist . . ." for if an atom radiated a classical spherical wave it could not recoil.

Einstein, A. "Zur Quantentheorie der Strahlung." Physika Zeitschrift *18 (1917): 121–128.*

# 135. Light Tweezer

Yes. A focused laser beam can exert a trapping force perpendicular to the beam direction of $2 \times 10^{-12}$ Newtons or more to keep cells confined in a microscope at the optical axis. The intensity gradient across the light beam is the source of the force.

In the simplest geometry, consider a semitransparent object with a diameter greater than the wavelength of the incident light but smaller in size than the diameter of the incident light beam. Let the light source be a parallel beam of light rays all of the same frequency, such as in a laser beam focused to the point $f$ by a symmetrical lens. The object tends to focus the light rays somewhat, changing the direction of the light rays. The sideways recoil of the object occurs to simply conserve the linear momentum. If the light beam has an intensity gradient, brighter in the center than near the edge, the object will receive a net push back toward the optical axis in the center. There must also be a recoil of the object in the direction of the original light beam, which usually is taken up by the apparatus and Earth because the object is on a horizontal platform. A one-celled paramecium remains well trapped in a microscope via this light tweezer technique, begun at Bell Labs in the 1970s.

When the object is smaller than the wavelength of the incident light, a more detailed analysis is required to understand the 3-D trapping and the quantum interference effects.

Optical tweezers have been widely used for several decades in applications as diverse as experiments on molecular motors in biology and the movement of Bose-Einstein condensates in physics. The capabilities of single optical tweezers have been greatly improved and extended by the development of tailored beams and by schemes for generating large numbers of trapping sites and shapes simultaneously.

Block, S. M. *"Making Light Work with Optical Tweezers."* Nature 360 (1992): 493–495.

Chu, S. *"Light Trapping of Neutral Particles."* Scientific American 266, no. 2 (1992): 70–76.

MacDonald, M. P., et al. *"Creation and Manipulation of Three-Dimensional Optically Trapped Structures."* Science 296 (2002): 1101–1103.

Ulanowski, Z., and I. K. Ludlow. *"Compact Optical Trapping Using a Diode Laser."* Measurement Science and Technology 11 (2000): 1778–1785.

# 136. Fluorescent Lights

Today, artificial illumination requires more than 25 percent of the electricity generated worldwide. There are two trends in "energy saving" technologies. The first trend is using improved lamps, such as fluorescent, mercury, sodium, metal halide, and halogen lamps. The second trend is improving the electronic circuit design for such lamps.

Although fluorescent lights are four to six times more efficient than incandescent lamps, there now exist many other types of light sources that are even more efficient. For the fluorescent lamp, its efficient production of the UV is extended into the visible by a powder coating inside the tube.

This powder absorbs the UV light and fluoresces in the visible. Very little heating of the fluorescent lamp occurs, so the efficiency occurs before the production of the visible light, with very little electrical energy being converted into thermal energy. The conversion process in the powder makes the tube useful for room lighting.

So why is the incandescent lamp so inefficient, converting only about 4 percent to 12 percent of the electrical energy to visible light? The incandescent lamp is simply a resistor whose filament temperature rises until it gets rid of thermal energy at the same rate that thermal energy is being generated in the filament. In a standard 100-watt, 120-volt bulb, the filament temperature is roughly 2550°C, about 4600°F, so that the thermal radiation from the filament includes a significant amount of visible light.

The output is 17.5 lumens per watt, compared to a maximum of 240 lumens per watt if all the energy could be converted to visible light. The reason for this poor efficiency is the fact that tungsten filaments radiate mostly infrared radiation at any temperature that they can withstand. An ideal thermal radiator produces visible light most efficiently at temperatures of about 6300°C (about 6600 K or 11,500°F). Even at this high temperature, a lot of the radiation is either infrared or ultraviolet, and the theoretical luminous efficiency is 95 lumens per watt.

Most fluorescent lights predominantly emit light in the visible part of the spectrum and they do emit some UV light, but only in a narrow range of the UV spectrum. Unfortunately, their UV emission range does not overlap the two small ranges of UV light needed by humans for the best functioning of certain internal organs, which receive some of the UV light that passes through the skin, as well as vitamin D production from 7-dehydrocholesterol in the skin.

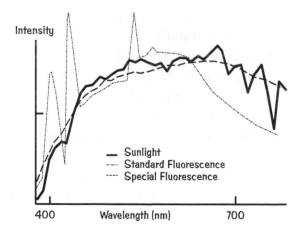

Special fluorescent lights more benevolent to human needs are available and mimic sunlight to produce a UV spectrum better matching the needs of these internal organs. Indeed, the lack of the required UV parts of the ambient light spectrum can lead to certain illnesses. Of course, the lack of

vitamin D production in the skin can lead to rickets and other problems associated with the calcium and inorganic phosphate metabolism. Eskimos and other indigenous peoples obtain plenty of vitamin D from the fish oils in their diets.

Porter, J. P., ed. How Things Work in Your Home (and What to Do When They Don't). New York: Henry Holt, 1985, p. 158.

# 137. Phase Conjugation Mirror

Yes, the light can return undisturbed if the light wave retraces it original path as its time-reversed twin *and* the medium retains its previous integrity. The phase conjugate of a wave possesses exactly the same spatial properties as the original wave, but it is said to be reversed in time. This means that a phase conjugate wave exactly retraces the path of the original beam. This method has the useful property that if a light beam propagates through a distorting medium, then the phase conjugate is produced and exactly retraces the path through the distorting medium, enabling the unfavorable effects of the distorting media to be reduced or eliminated. *Phase conjugation* is the general term for a process in which both the direction of propagation and the overall phase factor of a wave function are reversed.

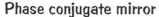

**Phase conjugate mirror**

**Conventional mirror**

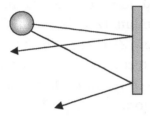

Some laser sources come with optical phaser conjugators to remove distortion in the laser beam. Optical phase conjugation occurs also when there are four waves mixing with all four waves of the same frequency. Another useful application of a phase conjugate mirror might be to put one in one reflecting path of an interferometer as a reference for detecting changes in the other path.

Blaauboer, M., D. Lenstra, and A. Lodder. "Giant Phase-Conjugate Reflection with a Normal Mirror in Front of an Optical Phase-Conjugator." Superlattices and Microstructures 23 (1998): 937.

Brignon, A., and J.-P. Huignard. Phase Conjugate Laser Optics. New York: John Wiley & Sons, 2004, chap. 1.

# 138. Stationary States

In the Bohr model of the hydrogen atom, one would calculate the frequency $f = 2\pi r/v$ of the electron's orbital motion. The virial theorem states that twice the kinetic energy plus the potential energy add to zero, so $mv^2 = ke^2 r$, from which the electron's frequency of orbit is $f = n^3h^3/(4\pi^2me^4)$. The actual Bohr energy $E = -2\pi^2me^4/(n^2h^2)$ is clearly a different quantity, and for an electron jump between two energy states, $E_2 - E_1 \neq hf_2 - hf_1$.

Spielberg, N., and B. D. Anderson. Seven Ideas That Shook the Universe, 2nd ed. New York: John Wiley & Sons, 1995, p. 248.

# 139. Angular Momentum

We take the space quantization of angular momentum as given, so there will be $(2j + 1)$ components in the z-direction from $j$ to $-j$, decreasing by an integer each step. Since there is no preferred direction, $J^2 = J_x^2 + J_y^2 + J_z^2$, that is, $J^2 = 3 <J_z^2>_{avg}$, where avg represents the average value given by $[j^2 + (j - 1)^2 + \ldots + (-j + 1)^2 + (-j)^2] h^2/(4\pi^2 [2j + 1])$. Using a math table or finding the sum of the series of squared integers directly, one can verify that $J^2 = j (j + 1) h^2/4\pi$.

Feynman, R. P., R. B. Leighton, and M. Sands. The Feynman Lectures on Physics. Vol. II, Reading, Mass.: Addison-Wesley, 1965, p. 34-11.

# 140. Kinetic Laser

The explosion of the lasing material creates many free electrons, some of which have been blown out of low-lying atomic states, creating the needed population inversion for possible lasing action. Practically any material can be used. During an extremely short time interval after the explosion—on the order of nanoseconds—stimulated emission may occur as photons from the exploding material exit the expanding blast volume. These photons pass through regions of the expanding cloud of ionized debris and can stimulate the emission of many more photons into the same quantum state at the same wavelength. The resulting coherent radiation at many frequencies, including the soft X-ray region, will show intensity spikes in particular directions.

Some of the first kinetic laser explosions were done at Livermore National Laboratory in the 1970s and 1980s with exploding foils and the Nova laser system. Since the first demonstration of soft X-ray lasing—emissions at about 10 nanometers or more—using the collisional excitation mechanism in neonlike selenium, many other neonlike ions, ranging from copper (Z = 29) to silver (Z = 47), have been made to lase. However, attempts to produce lower-Z neonlike X-ray lasers have been unsuccessful.

In the effort to develop a tabletop X-ray laser that would require smaller high-energy laser drivers than Nova and that could be used for applications such as biological imaging, nonlinear optics, holography, and so on, a prepulse technique has been developed. This technique has been used successfully to produce lasing in many lower-Z neonlike ions such as titanium (Z = 22), chromium (Z = 24), iron (Z = 26), and nickel (Z = 28). The use of this prepulse technique has opened up a new class of neonlike X-ray lasers for investigation.

Chapline, G., and L. Wood. "X-ray Lasers." Physics Today 6 (1975): 40.

Dunn, J., et al. "Demonstration of X-ray Amplification in Transient Gain Nickel-like Palladium Scheme." Physical Review Letters 80, no. 13 (1998): 2825–2828.

Nilsen, J. "Reminiscing about the Early Years of the X-ray Laser." Quantum Electronics 33, no. 1 (2003):1–2.

# 141. Noninversion Laser

Yes, lasing without inversion (LWI) can occur whenever absorption cancellation is established. Light amplification is then possible even when the upper-level population is less than the lower-level population. This cancellation can be set up in a three-level system in an atom in which the two absorption transitions to the same final state interfere and cancel, making the absorption probability zero.

In the diagram, upper-level state | a > is connected to lower levels | b > and | c >. Use incident photons of the

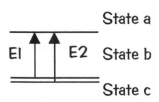

State a

E1 E2 State b

State c

**3-state system**

appropriate energies E1 and E2, which correspond to transitions | a > to | b > and | a > to | c >, respectively. The uncertainty in these atomic transitions leads to interference, since the transitions end in the same final state. There is no way to determine which absorption transition to the final state actually occurred, so like the Young double-slit experiment, one must have the interference. There is no interference between the emission paths, since they have different final states. By arranging the phases of the two incoming light rays properly, one can make the interference completely destructive for absorption. Then stimulated emission is the only process left. For details of the probability calculations, see the references below.

Narducci, L. M., H. M. Doss, P. Ru, M. O. Scully, S. Y. Zhu, and C. Keitel. " A Simple Model of a Laser without Inversion." Optics Communications 81 (1999): 379.

Scully, M. O., and M.S. Zubairy. Quantum Optics. Cambridge, Eng.: Cambridge University Press, 1997.

# 142. X-ray Paradox

The index of refraction $n$ for a material is normally stated with regard to the *phase velocity*, unless indicated otherwise. The phase velocity is $v_{ph} = c/n(k)$, where the index is a function of the wave number k. If $n(k) < 1$, then the phase velocity is greater than the speed of the light in the crystal. There is no alarm that the energy is being transported faster than c, for the group velocity is still less than c.

Essentially, travelling harmonic waves in all physical examples require wave packets or groups because of the non-infinite extent of space and/or time. There are two velocities associated with these wave packets or groups: the phase velocity and the group velocity. Harmonic waves or harmonic components have a phase velocity $v_{ph} = \omega/k$, where $\omega = 2\pi f$ and $f$ is the frequency. This phase velocity is the velocity at which the wave fronts travel. A group of harmonic waves or wave packet has a group velocity $v_g = d\omega/dk$, the velocity at which the packet shape or envelope travels—that is, the velocity at which information or energy is transported.

On the atomic level, the slowing of light passing through a material may be considered as a continuous process of absorption and emission of photons as they interact with the atoms of the material. One assumes that between each atom, the photons travel at c, as in a vacuum. As they impinge on the atoms, they are absorbed and nearly instantly re-emitted, creating a slight delay at each atom, which (on a large enough scale) seems to be an overall reduction in the speed of the photons. Quantum mechanically, the scattering is a two-step process of absorbing the incident photon and emitting a new photon.

Experiments in other ranges of the electromagnetic spectrum, particularly in the visible, have shown that by storing the phase information of the incident light beam in a gas vapor temporarily, one can even claim that the light pulse can be brought to rest!

*Addinall, E. "The Refractive Index of X-rays." Physics Education 6 (1971): 77–78.*

# 143. Benzene Ring

The benzene ring has six-fold rotational symmetry about an axis perpendicular to the plane of the ring. One simply requires a wave function solution of the Schrödinger wave equation that has this six-fold symmetry, and such a solution is easy to find. One would expect that knowing this solution would allow one to calculate the energy levels.

However, we are not done! There are two possible configuration base states, as shown in the diagram.

Both states should have the same energy, and they do. Therefore we really have a two-state system, analogous to the hydrogen molecular ion or the ammonia molecule, so the analysis should be for a two-state system. There will be the possibility that configuration A changes into configuration B. As a result, quantum mechanics will reveal that two new stationary states will occur, one state (the new ground state) with energy below the ground (lowest) state determined before, and one state with higher energy. The new ground state will be neither of the two configuration states shown but will be a linear combination of these two configuration states. Only this state is involved in the chemistry of benzene at room temperatures.

Understanding benzene was one of the first verifications of the linear superposition of states that is at the heart of quantum mechanics and also indicated that quantum mechanics will be successful at larger scales than atomic.

*Feynman, R. P., R. B. Leighton, and M. Sands. The Feynman Lectures on Physics, Vol III, Reading, Mass.: Addison-Wesley, 1965, pp. 10-10 to 10-12.*

# 144. Graphite

Place identical atoms into a diamond crystal structure. First, one would mathematically find a wave function for the four bonding electrons using the Schrödinger equation, resulting in what are called sp[3] orbitals. Then one would represent the periodic symmetry in the crystal. Each carbon atom will make four orthogonal bonds with tetrahedral symmetry if it can to its nearest neighbors. This diamond structure is one way to do this bonding.

Another way to have four carbon bonds is for six carbon atoms to form a regular hexagonal ring with two bonds in the ring for each carbon, and the other two bonds extending perpendicular to the ring, one upward and the other downward. Upon calculating the energy states for the four carbon binding states, one learns that the two perpendicular binding states are held less securely than the ones in the ring that form a plane. The structure makes graphite, a layered crystal that slips easily between the planes. Pencil writing surfaces have been made from graphite for several thousand years.

Carbon in the fullerene structure is even more interesting. The structure of 60 carbon atoms that results depends on many factors, including the velocity distribution of the free carbon atoms before collision, the formation of intermediate structures, and so on. Fullerenes tend to form by "rolling up" a graphite sheet and adding carbon pentagons to achieve curvature. If you just roll the sheet into a cylinder and cap off the ends with pentagon-curved hemispheres, you make a carbon nanotube. These nanotubes are quite different from the traditional fullerene-type materials (i.e., roundish cages), so they have quite different properties.

Collins, P. G., and P. Avouris. "Nanotubes for Electronics." Scientific American 283, no. 6 (2000): 62–69.

Pauling, L. General Chemistry. New York: Dover, 1988, pp. 168–170, 207–210.

# 145. Ozone Layer

Ozone plays two important roles with regard to the energy balance for Earth. As a minor greenhouse gas in all parts of the atmosphere, including near the surface, ozone helps maintain Earth's average temperature at about 13°C instead of a frigid –17°C. The concentration of ozone in the upper atmosphere, however, regulates the UV intensity in sunlight reaching the surface. All organisms need some UV light to maintain a healthy existence, but any reduction of ozone in the upper atmosphere might allow dangerously large amounts of UV to reach the surface.

The two polar regions are extremely

susceptible to ozone depletion, particularly by chlorofluorocarbon (CFC) molecules and others, because the ice crystals in the air provide these fluorocarbons with platforms for rapid ozone dissociation. Already, as a result of ozone depletions in the upper atmosphere above the polar regions, particularly above the South Polar region, there has been an increase in eye problems in land animals such as sheep in the southern parts of South America and in Australia and New Zealand.

*Allègre, C. J., and S. H. Schneider. "The Evolution of the Earth."* Scientific American *271, no. 4 (1994): 66–75.*

*Newchurch, M. J., et al. "Evidence for Slowdown in Stratospheric Ozone Loss: First Stage of Ozone Recovery."* Journal of Geophysical Research *108, no. D16 (2003): 4507.*

# 146. Greenhouse Gases

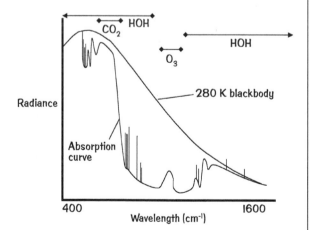

The greenhouse gases trap most of the infrared, and this additional energy helps heat Earth to its present average equilibrium temperature of about 13°C. Without the greenhouse effect in our atmosphere, Earth's average surface temperature would be about 256 K, or about –17°C, much too cold for many life forms. The greenhouse effect involves the influx of sunlight, its absorption by the atoms and molecules of the stuff on Earth, and the attempted emission of light and infrared energy back into space.

Although carbon dioxide receives the most attention in the press, HOH vapor is the most important greenhouse gas because the HOH molecule absorbs energy over practically the whole range of visible and infrared frequencies, while carbon dioxide absorbs in a small region of the near infrared only. Water vapor controls about 60 percent of the greenhouse effect, carbon dioxide about 20 percent, and the other trace gases in the atmosphere the remainder.

Additional greenhouse gas concentrations added to the atmosphere would be expected to trap even more infrared radiation and probably raise the temperature further. However, a convincing comprehensive model of this process has not been achieved. There are many complications to any model of Earth, including the transmission and reflection of light from clouds, the movements of the ocean currents, the effects of human-made sources and sinks, the perturbations by vegetation, land animals, and sea organisms such as plankton, the thermal energy input from additional heat sources such as mantle transport of thermal energy from the interior of Earth, and the effects of the bombardment by cosmic rays from the galaxy and beyond.

Many natural temperature records have been mined in the past few decades that provide the history of temperature changes, so fluctuations in average temperature, a vaguely defined concept, are not new. The claim seems to be that the *rate* of increase of the average temperature is among the greatest ever experienced on Earth. Whether this hypothesis is verified in the near future will take better models, meaning greater computing capability and more included physical and chemical processes and/or a definitive, unambiguous example.

Gillett, N. P., F. W. Zwiers, A. J. Weaver, and P. A. Stott. "Detection of Human Influence on Sea-Level Pressure." Nature 422 (2003): 292–294.

Herzog, H., B. Eliasson, and O. Kaarstad. "Capturing Greenhouse Gases." Scientific American 282, no. 2 (2000): 72–79.

# 147. LED vs. LCD

We assume that they all have the same resolution, and we know that all three types of display—LED, LCD, and plasma—require energy to operate. But the majority of the energy for the LCD is provided by the ambient light, whereas all the energy for the LED and plasma displays must be provided by the electronic power source itself, such as a battery or the AC supply. In addition, considerable thermal energy can be produced in a plasma display, an energy requirement beyond simply producing a picture on the screen. Of course, there are LCD displays that must provide their own ambient light if they are to be used in a dark environment, so these displays have additional energy requirements when operated in this manner.

So LCDs consume much less power than LED and gas-display models because LCDs work on the principle of blocking light rather than emitting it.

# 148. Sonoluminescence

The light produced by sonoluminescence must originate in atomic transitions, electrons in excited states in atoms jumping down to lower energy levels and emitting photons to conserve energy and angular momentum. The apparatus consists of distilled water with an admixture of a little helium or other inert gas in a spherical flask surrounded by a piezoelectric crystal or two to send in sound waves at practically any frequency. The details of the apparatus can be found at many sites on the Internet.

The sound energy creates bubbles in the water that rapidly collapse and emit a flash of light from their central region. Instead of sound waves, a powerful laser pulse also can create the bubbles for the pulse of light. The spectrum of the emitted sonoluminescent light pulse is similar to a blackbody spectrum of an object at about 8,000 K, hotter than the Sun's surface temperature of about 6000 K! And the pulse of light lasts for picoseconds, with such an intensity that it can be seen by the unaided human eye.

The reference below provides experimental results that support the popular theory that a plasma inside the bubble causes sonoluminescence. The research team fitted their pulses' spectra to a blackbody radiation curve and found the correspondence to plasma temperatures at about 8000 K. The gas in the bubble becomes a partially ionized plasma, and the radiation is emitted by an energy cascade from ions to electrons and finally to photons.

More details will be understood eventually as faster optical response systems become available to better follow the time development of the light emission process. In fact, how quickly a state-of-the-art photodetector system operates is measured against what initial parts of the sonoluminescent pulse of light can be discerned!

Baghdassarian, O., H.-C. Chu, B. Tabbert, and G. A. Williams. "Spectrum of Luminescence from Laser-Created Bubbles in Water." Physical Review Letters 86 (2001): 4934.

# 149. Siphoning Liquid Helium

At temperatures near absolute zero, normal liquid He I becomes superfluid liquid He II by undergoing a second-order phase transition. Its He atoms can move without viscosity in the superfluid. Superfluidity is a quantum mechanical phenomenon, with a macroscopic volume (centimeter dimensions) of liquid acting like a single macroscopic particle and described by a single-particle Schrödinger equation.

Immediately, superfluid He II in an open beaker will form a film that crawls up the walls, over the top, and down the sides until the beaker is emptied. Normal fluids also can be siphoned out of containers, but only if their motion is started externally! The solid surfaces in contact with He II are covered with a film 50 to 100 atoms thick along which frictionless flow of the liquid occurs. Supposedly, mass transport flow in the He II film takes place at a constant rate that depends only on temperature.

As the atoms of liquid He II move up the wall, they gain potential energy. What process provides the energy?

The answer lies in the ability of helium atoms to wet any surface—that is, normal liquid He I atoms cling to the wall. The helium-helium force is the weakest force in nature because the K shell of electrons is complete and the helium zero-point motion is significant, so the helium-anything force is stronger. Hence helium atoms would rather be next to anything other than another helium atom. So He atoms quickly form a film when presented with the wall of the container because the helium–anything attraction lowers the potential energy and so on, while they gain gravitational potential energy. These He atoms clinging to the wall are no longer in the superfluid phase because their flow velocities are now lower than a critical velocity value.

The thickness of the film is usually limited to a few hundred atomic diameters because at some thickness the advantage of being near to the wall is canceled by the increase in gravitational potential energy. Then, while the normal fluid is clamped to the wall, the superfluid He II flows freely as the He atoms on the wall act as a siphon.

Goodstein, D. L. States of Matter. Englewood Cliffs, N.J.: Prentice-Hall, 1975, p. 327.

# 150. Quantized Hall Effect

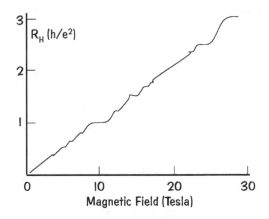

In a two-dimensional metal or semiconductor, the standard Hall effect is observed, but at low temperatures, a series of steps appear in the Hall resistance as a function of magnetic field instead of showing the typical monotonic increase. By confining the electron system in the third dimension to confine the electron gas to two dimensions, only specific electron wave functions meet the boundary conditions, so only certain quantized energy levels are available for the electrons. These steps in the Hall resistance occur at incredibly precise values of resistance, which are the same no matter what sample is investigated—that is, the resistance is quantized in units of $h/e^2$ divided by an integer. This amazing result is the quantized Hall effect.

Recall that electrons have a spin

1/2 and obey the Pauli exclusion principle. As electrons are added to an energy band, they fill the available energy band states, just as water fills a bucket. The states with the lowest energy are filled first, followed by the next higher ones. At absolute zero temperature T = 0 K, the energy levels are all filled up to a maximum energy called the Fermi level. At higher temperatures one finds that the transition region between completely filled states and completely empty states is gradual rather than abrupt and described by the Fermi function, which has a value of 1 for energies that are more than a few times $kT$ below the Fermi energy, equals 1/2 if the energy equals the Fermi energy, and decreases exponentially for energies that are a few times $kT$ larger than the Fermi energy.

Consider the ideal case of a fixed Fermi energy and a changing applied magnetic field. In the presence of the magnetic field, the density of electron energy states in 2-D is no longer constant as a function of energy and bunches into discrete energy levels, called Landau levels, of finite width separated by the cyclotron energy, with energy regions between the Landau levels where there are no allowed electron states. As the magnetic field is swept to higher values, the Landau levels move relative to the Fermi energy.

When the Fermi energy lies in a gap between Landau levels, there are no available states to scatter into, so there is no scattering, and the electrical resistance falls to zero. The Hall resistance for the Hall current cannot change from the quantized value whenever the Fermi energy is in a gap between Landau levels, so one measures a plateau. Only when the Fermi energy is in the Landau level can the Hall voltage change and a finite resistance value appear.

Kivelson, S., D.-H. Lee, and S.-C. Zhang. "Electrons in Flatland." Scientific American 274, no. 3 (1996): 86–91.

# 151. Integrated Circuits

Heat dissipation is the biggest problem in ICs. Simple, old-fashioned thermal energy limits the density of electronic components. The ability to miniaturize continues to improve, but unless thermal energy production per volume is decreased or new geometrical paths for thermal energy transport away from the sources are devised, the game is lost. At present, 3-D ICs offer a temporary reprieve, but even they will find their limit.

Some short time scale solutions may be possible. The best thermal conductor among crystalline materials is diamond, so going to a diamond substrate may be a solution. However, the technology of diamond is not yet competitive with silicon technology. Also, components on these substrates that require significantly less energy to function as a gate would delay the overwhelming impact of thermal problems. Optical information transfer between components would eliminate electrical currents and their thermal effects, but silicon does not have the right optical properties; hence the active research into doping silicon to make the desirable optical properties.

At the longer time scale of several decades, perhaps the silicon and semiconductor technology will simply fade away in favor of some other technology on the time horizon that seems unachievable today but that would become viable then. Or the newer technology hasn't even been dreamed about yet!

For any solid or liquid material, quantum disturbances from cosmic rays may decide the ultimate limit in electronic component density unless redundancy can solve this problem. For optical systems based on light interference, and so on, who knows what is possible? Whatever wins in the decades ahead will be numerous orders of magnitude smaller and faster, as well as more robust than what we have today.

# 152. Atomic Computers?

Yes. One can use electron spin directions as binary holders, for example. Even the nuclear spins can join in the game. Quantum computers already use nuclear spins for storage. On a bigger scale, DNA molecules are being used for a DNA computer.

Several difficulties in making atomic computers exist, but all of the difficulties can be overcome by clever techniques. Putting information in and reading information out of these atomic systems have been done in the laboratory already. Maintaining their fixed states is a different kind of problem that depends on which type of system is being used. Nuclear spin systems have been used quite successfully since the 1940s with the development of nuclear magnetic resonance (NMR). Electron spin systems also are controlled quite nicely in labs. If isolation of the system is required, then vacuum chambers work well for long enough time periods of particle isolation.

At the other extreme are proposed quantum computers utilizing the caffeine molecules in a cup of java. They are being bombarded constantly by the other molecules in the liquid, so the liquid environment brings about a rapid decoherence of the system. However, there are an awful lot of caffeine molecules in the cup, at least $10^{20}$. The quantum computer requires probably only a million or so to retain their isolation for the duration of the computation time—microseconds, perhaps—so the numbers may win out.

Like the limits to integrated circuit component density caused by thermal effects and by cosmic ray bombardment, atomic computers also may face similar limits. The type of atomic computer devised will determine how hostile the environment can be.

## 153. X-ray Laser?

The mechanism for the intense X-ray source appears to be the following according to K. DasGupta, the originator of this unique X-ray source. The W X-rays from the Cu-W X-ray tube knock out K shell electrons and others in the Cu atoms in the external Cu crystal to produce a temporary (about $10^{-15}$ second) population inversion, and the Cu X-rays coming simultaneously from the same tube then stimulate transitions in these Cu atoms to produce the Cu $K\alpha_1$ line at the Bragg angle to the Cu(111) atomic planes. This mechanism is very selective, the line being so narrow and intense and the process being so efficient that one does not detect any of the competing Cu $K\alpha_2$ emission to the available 1s state. The single-frequency intense X-ray line has been used to analyze materials in minutes that formerly required hours to days to accumulate enough data.

Whether the population inversion for the 2p–1s transition in the external Cu atoms actually occurs is unknown. The emission X-ray line is uncharacteristically narrow and intense, and the absence of the other competing line indicates that whatever the selection process is doing must be very efficient. Other element sources such as nickel, based on the same mechanism, also have been made.

DasGupta, K. "CuKa1 X-ray Laser." Physics Letters A 189 (1994): 91–93.

## 154. Bose-Einstein Condensate

A Bose-Einstein condensate is formed at the coldest temperatures, which means that the atoms have been slowed in their motion to be almost stationary. By the de Broglie relation, each atom of mass $m$ has a de Broglie wavelength $\lambda = h/p$, where $p$ is its momentum $mv$ and h is Planck's constant. As the velocity $v$ is further reduced to cool the atoms, the de Broglie wavelength increases accordingly. Eventually temperatures are reached for which the wavelengths of adjacent and nearby atoms begin to overlap in space considerably. Further cooling places all the atoms in intimate contact in one collective quantum state. Individual atoms can no longer be discerned because they act like one big "atom."

The first Bose-Einstein condensate was achieved in 1995, even though the physics principles have been known since Einstein and Bose proposed them in the 1920s. About 2000 rubidium atoms in the gas were cooled to 170 nanoK when they formed a Bose-Einstein condensate less than 100 micrometers across. The condensate lasted for about 15 seconds and was cooled further, all the way down to 20 nanoK.

Anderson, M. H., et al. "Observations of Bose-Einstein Condensation in a Dilute Atomic Vapor." Science 269 (1995): 198.

Castin, Y., R. Dum, and A. Sinatra. "Bose Condensate Make Quantum Leaps and Bounds." Physics World (August 1999): 37.

Cornell, E. A., and C. E. Wieman. "The Bose-Einstein Condensate." Scientific American 278, no. 3 (1998): 40–45.

Townsend, C., W. Ketterle, and S. Stringari. "Bose-Einstein Condensation." Physics World (March 1997): 29–34.

# 155. Quantum Dots

Quantum dots are crystals, essentially metal or semiconductor boxes, containing only a few hundred atoms and a well-defined number of electrons. The number of electrons can be controlled by the electrostatic environment. The trick is to adjust how many electrons end up in each quantum dot.

An electron in a 3-D box is constrained to have a quantum mechanical wave function that matches the boundary conditions for the Schrödinger wave equation, producing discrete energy levels that are inversely proportional to the square of the wavelength. As the box is made smaller, the energy levels become farther apart. If the quantum dot diameter—that is, box diameter—is made small enough in fabrication, only a few energy levels will exist inside for the electron. Hence one can make quantum dots small enough to allow only one fluorescence transition possible in the visible part of the spectrum.

The data from the first quantum dot spectrum showed a rich harmonic series of transitions between electron energy levels. Subsequent tweaking of the electrostatic potential was shown to reduce the dot size and increase the energy spacings. Later researchers have been able to magnetically link together quantum dots with the hope of making arrays of them for quantum computing.

Flügge, S. "Particle Enclosed in a Sphere." In Practical Quantum Mechanics. Vol. I. New York: Springer-Verlag, 1974, pp. 155–159.

Reed, M. A. "Quantum Dots." Scientific American 268, no. 1 (1993): 118–123.

Whitesides, G. M., and J. C. Love. "The Art of Building Small." Scientific American 285, no. 3 (2001): 38–47.

# Chapter 8
# Chances Are

## 156. Schizophrenic Playing Card

According to the rules of QM, the final state should be the superposition of the two alternative falling directions, with equal amplitudes $\psi_1$ for left and $\psi_2$ for right. But we never see a card fall both ways simultaneously. Any air molecule colliding with the card is equivalent to an observation, a measurement process, so QM rule 3 applies and the outcome reduces to the classical one, with equal probabilities $P_1$ to fall to the left side and $P_2$ to fall to the right side.

The term describing this reduction of the wave function to the classical probabilities that have no QM interference is often called decoherence. The Schrödinger equation, which is deterministic, controls the entire process.

*Tegmark, M., and J. A. Wheeler. "100 Years of Quantum Mysteries." Scientific American 284, no. 2 (2001): 68–75.*

## 157. Schrödinger's Cat

In QM, it is irrelevant whether you actually peek or not. If *in principle* you could have determined the status of the cat, QM reduces to the classical result by rule 3. The cat is now either alive *or* dead, not both. The two QM alternatives reduce to just one possibility.

Note that this example with the cat brings the connection between the non-intuitive behavior of Nature on the microscopic scale up to the macroscopic scale of our everyday experiences. There has been an enormous amount of controversy over this example and its interpretation. Some of the issues are discussed in the references listed below.

*Albert, D. Z. "Bohm's Alternative to Quantum Mechanics." Scientific American 277, no. 5 (1994): 58–67.*

*Loeser, J. G. "Three Perspectives on Schrödinger's Cat." American Journal of Physics 52 (1984): 1089–1093.*

*Wick, D. The Infamous Boundary: Seven Decades of Heresy in Quantum Physics. New York: Copericus Books, 1996, pp. 149–152.*

*Yam, P. "Bringing Schrödinger's Cat to Life." Scientific American 276, no. 6 (1997): 124–129.*

## 158. Wave Functions

No. Beyond three dimensions there is no direct one-to-one correspondence between many-dimensional configuration space coordinates and the three-dimensional coordinates of position space.

The misconception referred to here shows up in discussing the wave function for two-particle systems, especially when the discussion refers to the two-particle wave function reducing to the classical result. One often encounters questions about how the wave function can reduce instantaneously to the result, as if there has been some faster-than-light information transfer. Fortunately, the two-particle wave function reduces in configuration space, not in position space!

Hilgevoord, J. "Time in Quantum Mechanics." American Journal of Physics 70 (2002): 301–306.

Mermin, N. D. "Is the Moon There When Nobody Looks? Reality and the Quantum Theory." Physics Today 38, no. 4 (1985): 38–47.

Styer, D. F. "Common Misconceptions Regarding Quantum Mechanics." American Journal of Physics 64 (1996): 31–34.

Wick, D. The Infamous Boundary: Seven Decades of Heresy in Quantum Physics. New York: Copericus Books, 1996, pp. 162–166.

# 159. Wave Function Collapse?

The original wave function $\Psi = \psi_1 + \psi_2 + \psi_3 + \ldots$ will change. The probe photon did not scatter off the electron in particular imaginary boxes, so we know immediately that the wave function should not include their amplitudes. One could say that there has been a partial collapse of the wave function even though there has been no interaction! We believe that this gedanken experiment was discussed first by physicist Robert H. Dicke in the reference below.

Dicke, R. H. "Interaction-Free Quantum Measurements: A Paradox?" American Journal of Physics 49 (1981): 925–930.

Hilgevoord, J. "Time in Quantum Mechanics." American Journal of Physics 70 (2002): 301–306.

# 160. Quantum Computer

A quantum computer relies on maintaining its linear superposition of quantum states—that is, $\Psi = \psi_1 + \psi_2 + \psi_3$, its coherence during the calculations—so that all the states participate in the calculation. Quantum decoherence is a bad thing for a quantum computer. A collision with the wall of the chamber or with another molecule will ruin the coherence because an observation has been made. By QM rule 3, we no longer sum over the amplitudes $\psi_i$. This decoherence then ruins the quantum computation because only one state will be participating in the computations.

Maintaining coherence in a real physical system has been progressing slowly for the past decade, with coherence times of tens of nanoseconds for three identical subsystems working as a quantum computer. No one knows what type of physical system will compose the first 18-subsystem quantum computer in the future, but this computer probably will outdo all the other classical computers combined in computing speed.

Awschalom, D. D., M. E. Flatté, and N. Samarth. "Spintronics." Scientific American 286, no. 6 (2002): 67–73.

Lloyd, S. "Quantum-Mechanical Computers." Scientific American 273, no. 4 (1995): 140–145.

Nielsen, M. A. "Rules for a Complex Quantum World." Scientific American 287, no. 5 (2002): 67–75.

## 161. Cup of Java Quantum Computer

Coffee contains caffeine molecules, which may be useful as quantum subsystems for a quantum computer because they contain two rings in a plane with many attached hydrogen atoms. The nuclear spin states of the H atoms attached to the rings can be used for information storage à la NMR. That is, a nuclear magnetic resonance (NMR) system is a collection of nuclear spin states in an external magnetic field that tend to align the spins. In the simplest ideal case at temperature $T$, the external magnetic field $B$ is uniform and there are two spin states, up and down. Let's say that $B$ aligns most of the spins to the up state, with the ratio of down to up spins being determined by the exponential factor $\text{Exp}(-\mu B/kT)$, where $\mu$ is the nuclear magnetic moment and k is the Boltzmann constant. An external radiofrequency pulse of the proper frequency $v$ and energy $hv = 2\mu B$ can flip a down spin to an up spin for a stimulated absorption transition or can cause a stimulated emission of a photon by a spin flip from up to down.

Now for some coffee. The liquid contains about $10^{20}$ caffeine molecules. Even if we assume that all of them participate initially in bunches as coherent states of many quantum computers in the cup just before the calculation, most bunches will experience collisions during the calculation time of a nanosecond, say, and drop out from the collection of coherent states of the system. However, a significant number of bunches of coherent states may be participating still when the calculations are done, and these will provide a strong signal above the background noise. At least that's the hope!

Gershenfeld, N., and I. Chuang. "Bulk Spin Resonance Quantum Computation." Science 275 (January 17, 1997): 350.

## 162. Bragg Scattering of X-rays

Bragg scattering requires $\lambda < d$; therefore there will not be any collective scattering from a group of scatterers at *different* atoms within one wavelength. The actual scatterers of the X-rays are the electrons at each atom in these planes of the crystal. Coherent scattering requires *fixed* phase relationships, but there is no *fixed* phase relationship between electrons at different atoms nor between the electrons doing the scattering at any moment. Therefore, the X-rays scattered into the Bragg angle have a multitude of *random* phases and not fixed-phase relationships. The scattering probability is proportional to $N$, the number of scatterers, and not $N^2$, as it would be for coherent scattering.

Here is the QM argument mathematically. Let $\psi_i$ represent the probability amplitude to scatter an X-ray at the $i^{th}$ atom. We know from QM rule 2 that $\Psi = \psi_1 + \psi_2 + \psi_3 + \ldots$, for alternative ways to go from the X-ray source to the crystal to the X-ray detector. Each $\psi_i$ represents one atom, and we assume single scatterings on the way to the detector for simplicity. Each $\psi_i = \exp[i\delta]\, \phi_i$, which includes a phase part $\exp[i\delta]$ and the identical scattering amplitude $\phi_i$ at the identical atoms in the crystal. If the phase part at each scattering atom is identical, then we would have $\Psi = N \psi_1$ and the probability $P = N^2 |\psi_1|^2$, giving us coherent scattering proportional to $N^2$.

However, there is no correlated motion between electrons on different atoms, so their phases are random. If the phase differences between scatterers—that is, the electrons on different atoms—are not fixed differences, then the sum is over random phases and, like the random walk problem, the

total amount is proportional to $\sqrt{N}$ instead of $N$. Therefore $\Psi = \sqrt{N}\, \psi_1$, so $P = N |\psi_1|^2$. The Bragg scattering of X-rays is not a coherent scattering process.

# 163. Beautiful Faces

Coherent scattering of light by the atoms in the skin is the reason for our ability to see details of a face. The ambient incident light is scattered by the molecules of the skin. Two factors are significant for this two-step scattering process: the time interval required and the number of coherent scatterers. In the visible region of the electromagnetic spectrum, this scattering process occurs in atoms in less than $10^{-8}$ second over an area of the skin involving about a million atoms within a circle with a radius of about one wavelength of the light. The wavelength of greenish light is about 500 nanometers.

Consider scattering one incident photon at a time. During the scattering time of a single photon by these one million alternative paths there is almost no movement of the scattering atoms in the molecules, so alternative paths have essentially fixed phase relationships. By QM rule 2, $\psi = \psi_1 + \psi_2 + \psi_3 + \ldots$, and $\psi = N \psi_1$ with probability $P = N^2 |\psi_1|^2$, giving us coherent scattering proportional to $N^2$. With incoherent scattering we would not see much detail.

In the UV, both factors are smaller than for light in the visible spectrum—the scattering occurs in less time, and the area for each scattering is less and involves fewer atoms because the wavelength is much less. The face seen in the UV would appear grainier with less detail because the adjacent coherent scattering areas are smaller and the shorter time interval means that they will have some effects of almost-random phases.

In the IR, most of the scattering involves molecular transitions, which are relatively slow processes, so the scattering process involves a much longer time interval. But each molecule itself is completely involved in the scattering. So even though the wavelength is large, involving many more scattering centers, the molecular scatterers move significantly during the IR scattering process, producing random phases everywhere and a smearing of the image.

Organisms of many different types see in the UV and/or in the IR to find their nourishment, as well as in the visible. However, we humans evolved without being able to see either the UV or the IR, our vision being confined to the visible part of the electromagnetic spectrum. Why our eye-brain system evolved in this way is not known.

# 164. Gravitational Waves

Yes, the coherent scattering of gravitational waves is expected to occur, with the scatterers being mass quadrupoles—that is, mass pairs in the antenna. J. Weber, the same physicist who first calculated the classical cross section for gravitational wave scattering in 1959, proposed in 1981 that the coherent scattering of gravitational waves would enhance the scattering cross section for certain detectors by a factor of $10^6$ or more. The larger cross section might explain the large responses of his two independent one-ton cylindrical aluminum bar gravitational wave detectors every time either end faced the nucleus of the Milky Way galaxy, approximately twice per day. If his proposal for a coherent scattering response is correct, then solid bar antennas would be much more sensitive to gravitational waves than large interferometers with their small masses at the mirrors such as LIGO and VIRGO.

The QM calculation can be outlined as follows. With wavelengths in the kilometer range being much longer than the size of the Al bar antenna in the lab, all the mass pair quadrupoles in the antenna are within this one wavelength. Hence, their responses are approximately in phase, and each mass pair offers an equivalent alternative scattering path. By QM rule 2, $\Psi = \psi_1 + \psi_2 + \psi_3 + \ldots$, and $\Psi \sim N \psi_1$ with probability $P = N^2 |\psi_1|^2$, giving us coherent scattering proportional to $N^2$, where N is the total number of mass pairs in the bar, about $10^{24}$. However, the bar is actually composed of many microcrystallites, so one really sums the QM amplitudes over the number of mass pairs within each microcrystallite, then sums the probabilities over all the microcrystallites. The coherent scattering probability is still more than 10 million times larger (after accounting for the crystalline defects) than the classical non-coherent scattering response that Weber first calculated in 1959.

Whether any bar antenna for gravitational waves behaves as a coherent scatterer has not been unambiguously demonstrated. Instead of the classical result with the bar oscillating at its resonant frequency and its harmonics when hit by a pulse of gravitational waves, the coherent scattering bar would essentially have an almost equal response to a wide range of frequencies. The actual experimental bar responses are complicated and require elaborate methods to find gravitational wave scattering signals buried in background noise.

If the Weber bars were really detecting gravitational waves from the galactic nucleus, there is an enigma when the original classical response cross section is used. The rate of conversion of mass to energy at the galactic nucleus should have devoured the whole galaxy by now! I suppose that we must wait for LIGO and VIRGO to detect and calibrate gravitational waves before we truly know whether gravitational waves can scatter coherently in Weber bar antennas.

Gibbs, W. W. "Ripples in Spacetime." Scientific American 286, no. 4 (2002): 62–71.

Preparata, G. "Superradiance Effect in a Gravitational Antenna." Modern Physics Letters 5 (1990): 1–5.

Weber, J. "Gravitons, Neutrinos, and Antineutrinos." Foundations of Physics 14 (1984): 1185–1209.

# 165. Coherent Neutrino Scattering

In 1984, so the story goes, J. Weber proposed to build a detector for the coherent scattering of neutrinos in a proposal for research monies. The proposal review committee challenged him to write up the neutrino coherent scattering idea and publish the paper in a reputable physics journal. In December 1984 he submitted the paper, "Method for Observation of Neutrinos and Antineutrinos," to *Physical Review C,* and the paper was accepted by a referee within eight days of the December 12 reception date!

The paper triggered an enormous

response in parts of the physics community. Numerous rebuttals of his arguments appeared in the physics literature within months after this publication, but all of these rebuttals can be refuted. Every paper erroneously assumes that the nuclear scatterers act as potentials. Wrong! Weber shows in the first section of the paper that such an assumption *cannot* lead to coherent scattering for neutrino wavelengths less than the spacing between nuclei. However, everyone seems to ignore the details presented by Weber, who correctly explains why the nonrelativistic calculation does not predict coherent neutrino scattering for neutrino wavelengths less than the atomic spacing. The QM argument is essentially dependent on the fact that the scattering phases among the nuclei will be random, leading to a scattering probability proportional to $N$ instead of $N^2$.

In later parts of the paper Weber does the relativistic QM scattering calculations to show that coherent scattering for all energies occurs—that is, neutrinos of all energies will suffer coherent scattering. Included in these calculations are terms involving the stiffness of the defect-free crystal, and so on. The conceptual idea is that when the crystal as a whole recoils, like a Mössbauer Effect scattering, then one cannot determine (even in principle) where the nuclear scattering

of the neutrino took place. Hence their responses are in phase and offer equivalent alternative scattering paths. One must sum the amplitudes over all possible paths—that is, all nuclei—to obtain the total amplitude for the neutrino scattering.

By QM rule 2, $\Psi = \psi_1 + \psi_2 + \psi_3 + \ldots$, and $\Psi = N \psi_1$ with probability $P = N^2 |\psi_1|^2$, giving us coherent scattering proportional to $N^2$, where $N$ is the total number of nuclei in the bar, about $10^{23}$. One gains the enormous factor of $10^{23}$ for neutrino scattering over the noncoherent cross section! The only remaining contention is whether all the phase relationships are properly accounted for in this relativistic calculation.

Weber (now deceased) actually conducted several experiments to check his relativistic calculations for a long defect-free single crystal detector. He claims to have verified the turning on and the turning off of a nuclear reactor in blind tests, the leaking of tritium from a highly radioactive tritium source, and the twice-daily passing of the Sun though the long axis of his crystal detector. In 1995 he determined that the total measured solar flux of neutrinos—all three types, because the detector did not distinguish among them—was equal to the total neutrino flux expected by the standard solar model. This predicted result agrees with the 2002 results reported by the heavy water detector at the Sudbury Neutrino Observatory (SNO).

Ho, T. H. *"Comments on the 'Method for Observation of Neutrinos and Antineutrinos.'"* Physics Letters 168B (1986): 295.

Weber, J. *"Method for Observation of Neutrinos and Antineutrinos."* Physical Review C 31 (1985): 1468–1475.

# 166. Magnetic Resonance Imaging (MRI)

Nuclear magnetic resonance experiments began in the 1940s, and they continue to be very useful today. Their alternative QM behavior is described as a collection of spins acting together. Initially, the spin collection has a total spin $S$ in a collective quantum state $\Psi = \psi_1 + \psi_2 + \psi_3 + \ldots$ and then the pulsed magnetic field rotates them all so slightly to $S - \alpha$ with respect to the original direction—that is, they act collectively and coherently. No one spin behavior is isolated from the others in the same microscopic atomic environment. All hydrogen nuclei in the same environment respond the same, while those in a different environment respond slightly differently.

The MRI instrument for magnetic resonance imaging uses the differences in the microscopic atomic environment to allow different regions of the living tissue to be "seen" separately. A computer algorithm analyzes the data from numerous RF detectors surrounding the body and constructs an artificial image on a display screen. A dynamic MRI instrument has a fast response time to show changes occurring in the microscopic environment in seconds or less, such as muscle action or heart contractions.

# 167. Heisenberg Uncertainty

The uncertainty principle places no limit to the accuracy of measuring the particle's position. The uncertainty principle $\Delta p_x \Delta x \geq h/4\pi$ forbids the *simultaneous* measurement of both position and momentum in the same direction to arbitrary accuracy, not an individual measurement. Of course, practical design limitations exist that probably limit the measurement, but conceptually there is no limit. The same argument applies separately to the momentum.

An application of the Heisenberg uncertainty principle to the hydrogen atom is an insightful example. The hydrogen atom is usually solved in spherical polar coordinates instead of Cartesian coordinates. In spherical polar coordinates, the uncertainty relations are a bit more complicated and the consequences can be somewhat bizarre. For example, since the hydrogen wave function for the electron about the z-axis—that is, in the φ direction—is known precisely for the 1s atomic state, and hence the angular momentum has no uncertainty, the uncertainty in φ is maximum. Therefore, in the φ direction, one finds an equal probability at all angles, producing the smeared-out probability distribution in φ.

**Two waves added**

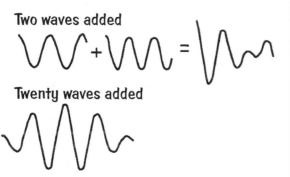

**Twenty waves added**

Many other uses for the uncertainty relation exist because it lies at the very heart of quantum mechanics. However, one can see that any description of a phenomenon using waves of any kind will require an uncertainty relation. Engineers are familiar with the fact that about a one-MHz bandwidth is required to reproduce a one-microsecond pulse: $\Delta f \Delta t \sim 1$, for example. Suppose there is a single-frequency wave defined by $y = y_1 \sin k_1 x$. This wave extends from $-\infty$ to $+\infty$, and the question "where is the wave located?" has no answer. By adding

together many single-frequency waves of different frequencies with properly chosen amplitudes and phrases, we can build up a lump in a narrow region of space of approximate length $\Delta x$. The range of wavelengths $\Delta \lambda$ needed can be represented by the corresponding range of wavenumbers $\Delta k$. The approximate mathematical relationship $\Delta x \Delta k \sim 1$ can be established by considering several examples, as seen in the Krane reference below.

Bohr's famous measurement disturbance argument is faulty. For half a century physicists have regurgitated this argument of how the uncertainty principle acts to defend quantum theory. In experiments that first refuted Bohr's argument, a beam of cold rubidium atoms is split to travel along two different paths; call them A and B. The beams still overlap and combine at the end of their journeys to create an interference pattern. Now the researchers looked to see which path the atoms followed by tweaking those on path B into a higher energy state by a pulse of microwaves. These atoms in

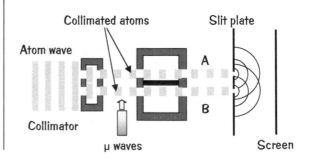

their internal states kept a record of which path they had taken. The microwave pulse absorbed by an atom is insignificant by a factor of about 10,000 and can cause little change to the atom's momentum, not enough to smear the interference pattern. Yes, QM worked still. With the microwaves off, interference fringes appear. Turn them on so you can tell which path was taken, and the interference pattern vanishes. The uncertainty principle is correct still, but the argument that "measurement disturbs the system" to explain the double slit experiment is wrong.

So what may be the deeper mechanism at work in the double slit experiment, for example? Perhaps quantum entanglement, in which every particle is linked to every other particle it has interacted with. Two-particle wave functions are linked in a six-dimensional configuration space with no one-to-one correspondence to physical 3-D space, so the entanglement of N particles will be described by a wave function in 3N-dimensional configuration space with no one-to-one correspondence to 3-D physical space. And now the mathematics becomes messier!

Dürr, S., and G. Rempe. "Can Wave-Particle Duality Be Based on the Uncertainty Relation?" American Journal of Physics 68 (2000): 1021–1024.

Englert, B.-G., M. O. Scully, and H. Walther. "The Duality in Matter and Light." Scientific American 271, no. 6 (1994): 86–92.

Krane, K. Modern Physics, 2nd ed. New York: John Wiley & Sons, 1995, pp. 93–106.

Styer, D. F. "Common Misconceptions Regarding Quantum Mechanics." American Journal of Physics 64 (1996): 31–34.

Wick, D. The Infamous Boundary: Seven Decades of Heresy in Quantum Physics. New York: Copericus Books, 1996, pp. 152–156.

# 168. Vacuum Energy?

There is always the zero-point energy in the vacuum. Whatever QM model for the vacuum is considered, all can be reduced in a first approximation to a large number of harmonic oscillators, which have a zero-point energy value that is non-zero. At present, QM calculations of the energy density of the vacuum seem to be too large by at least 30 orders of magnitude! The vacuum energy density should be about $10^{-11}$ J $m^{-3}$ if this vacuum energy is the source of the accelerated expansion of the universe determined by the Type 1a Supernova measurements in 1998.

One can do an energy estimate using the Heisenberg uncertainty principle. Or, if the vacuum has an effective potential for a scalar field, the product of the visible matter density and the potential will give the energy density for an assumed radius of the universe. In either case, the assumptions necessary to estimate this energy density would take us too far astray.

However, we can determine whether an electrically neutral particle of mass $\Delta m$ popping into existence for a time interval $\Delta t$ can be detected by its gravitational field. We use the uncertainty relation $\Delta E \Delta t \geq h/4\pi$ in the form $c^2 \Delta m \Delta t > h/4\pi$. Suppose we have the most sensitive detector, a free particle of mass $M$ initially at a distance $R$ away from $\Delta m$; then in the Newtonian approximation the detector will receive a pulse $P = F\Delta t$. Substituting $F = GM\Delta m/R^2$ into the uncertainty relation produces $GM\Delta m\Delta t/R^2 \geq GMh/(4\pi R^2 c^2)$. The initial state of the detector also obeys the uncertainty relation $\Delta P \Delta X \geq h/4\pi$, so that $\Delta m$ to be noticeable requires the impulse $P$ to be greater than about $2\Delta P$, or $\Delta X \geq 4R$ $(R/r_g)$, where the Schwarzschild radius of the detector $r_g = GM/c^2$. For objects ranging in size from protons to planets, $r_g$ lies within the object itself. So the momentum transferred by the impulse will not be detected!

Haroche, S., and J.-M. Raimond. "Cavity Quantum Electrodynamics." Scientific American 268, no. 4 (1993): 54–62.

Ostriker, J. P., and P. J. Steinhardt. "The Quintessential Universe." Scientific American 284, no. 1 (2001): 46–53.

Stefanski, B. Jr., and D. Bedford. "Vacuum Gravity." American Journal of Physics 62 (1994): 638–639.

# 169. Casimir Effect

Although the classical vacuum is a void, the quantum mechanical vacuum is a soup of virtual particle-antiparticle pairs that interact with the real atoms in the metal plates, these pairs being created and annihilated in extremely short time intervals in accordance with the Heisenberg uncertainty principle. That is, the more the total energy $\Delta E$ in the pair, the less time duration $\Delta t$ is its existence so that $\Delta E \Delta t \geq h/4\pi$. This vacuum pair "soup" pushes inward at both plates when the plates are very close to each other because certain particle-antiparticle pairs are practically forbidden from momentarily appearing between them. Essentially, if their deBroglie wavelength exceeds the plate spacing, these pairs have a much lower probability to be between the plates. But these same pairs appear outside the plates and provide the additional forces, whence the net inward force. Known as the Casimir effect, it was first measured in 1958.

The Casimir force is too small to be observed for plates that are not within microns of each other. Two mirrors with an area of 1 cm$^2$ separated by a distance of about 1 µm have an attractive Casimir force of about $10^{-7}$ N. Although this force seems very small, at distances of less than a micrometer the Casimir force becomes the strongest force between two neutral objects! At separations of 10 nanometer—roughly 100 times the size of an atom—the Casimir effect produces a force that is the equivalent of 1 atmosphere of pressure. The resurgence of interest in the Casimir force is because micromechanical devices on the scale of tens of nanometers must accommodate its effects!

Haroche, S., and J.-M. Raimond. "Cavity Quantum Electrodynamics." Scientific American 268, no. 4 (1993): 54–62.

Kleppner, D. "With Apologies to Casimir." Physics Today 43, no. 10 (1990): 9–11.

# 170. Squeezing Light

Classically, a ray of light is an electromagnetic wave having an amplitude and a phase, both being expressed in terms of the electric field components $E_x$ and $E_y$. Quantum mechanically, the normal modes of the electromagnetic field are quantized and treated as an ensemble of harmonic oscillators, one harmonic oscillator per normal mode. The number of photons in each harmonic oscillator is the energy in the corresponding oscillator. An harmonic oscillator obeys the Heisenberg uncertainty principle, so one expects the electromagnetic field to behave likewise.

As the electric field in a light ray is reduced, even a ray from a laser source, the fixed amount of intrinsic quantum noise in the light intensity becomes more obvious. This quantum noise in an electrical field is ever present. If you shine any light on a photodectector such as a photodiode, there will be fluctuations in the diode current corresponding to the individual photons being detected. One sees that the photons are not spread out evenly in time nor in spatial extent. Heisenberg's uncertainty relation dictates this behavior. The QM operators of phase- and amplitude-quadrature (i.e., for the perpendicular components of the $E$ field) of the electromagnetic field do not commute, similar to position and momentum of a particle. The product of phase- and amplitude-uncertainty has a fixed lower limit. The more precisely the phase of a light wave is measured, the less determined becomes its amplitude and vice versa. States of the light with the smallest possible amount of overall quantum noise are minimum uncertainty states.

The reduction in quantum noise in one observable of the light (e.g., the phase) at the expense of enhancing it in the complementary observable (i.e., the amplitude) can be done by parametric amplification procedures. The resulting states of the light are called *squeezed states,* since the quantum noise got squeezed at a particular phase angle. Their wave packets oscillate in time and get wider and narrower—that is, they breathe.

Alternately, the uncertainty in the amplitude of a laser beam can be reduced to a level below that normally allowed by the Heisenberg uncertainty principle, a level known as the zero-point quantum noise level. However, this increased knowledge comes at the expense of greater uncertainty in the frequency of the light. Essentially, one is using an uncertainty relation of the form $\Delta E_x \Delta E_y \geq V$, where V is a constant. Reducing the uncertainty in $E_x$ to $gE_x$ means that the uncertainty in $E_y$ becomes $E_y/g$ to keep their product the same.

Experiments with squeezed light promise to enhance our understandings of quantum mechanics at the individual atom and photon levels. Recently, a new type of ultraprecise laser pointer made by "squeezing" a beam in two directions was able to position the beam with a precision of 1.6 Å, about 1.5 times better than the theoretical limit for a conventional laser.

*Treps, N., et al. "A Quantum Laser Pointer." Science 301 (2003): 940–943.*

# 171. Electron Spin

Yes. Although the vacuum influence on the electron spin is extremely small, the same effect of the vacuum on the muon's spin has been measured at Brookhaven National Laboratory. The interaction magnitude is predicted by the Standard Model (SM) of Leptons and Quarks and their interactions. All fundamental particle-antiparticle pairs momentarily appear in the vacuum and disappear sporadically, so the electron (and muon) see them all, if only for a fleeting moment. This vacuum "soup" is slightly magnetic, so it increases the magnetic moment of the electron or muon to $g = 2(1 + a)$. The small correction of about 0.12 percent is called the anomalous moment but is often referred to as "g-2." Its measurement with gradually increasing accuracy presents spectacular agreement with calculation to better than 24 parts per billion.

The muon is 206 times heavier than the electron, so the muon's magnetic moment is 206 times smaller, but the virtual particles in the quantum soup can be more massive. As a result, the anomalous moment is 40,000 times more sensitive to undiscovered particles and new physics at short distances. There is agreement to 4 parts per million that must be regarded as the best test of the theory, but there is also a small discrepancy that needs to be explained, a difference in mean values of the experiments and the theory by 2.6 standard deviations.

The muon g-2 result cannot at present be explained by the established SM. Recalculations of the predicted theoretical value continue, and corrections have been done. Moreover, the g-2 calculation involves three of the four fundamental interactions—weak, electromagnetic, and color—so there are many Feynman diagrams that contribute.

Perhaps this unresolved g-2 difference is the harbinger of new physics beyond the SM, such as new quarks, or supersymmetric particles, or a surprise in the vacuum.

*Bennett, G. W., et. al. "Measurement of the Positive Muon Anomalous Magnetic Moment to 0.7 ppm."* Physical Review Letters 89 (2002): 101804.

## 172. Superconductivity

The paired electrons in superconductors that are in the superconducting state show Bose-Einstein condensation to a single macrostate. There is some small energy width to this macrostate because the pairs are composed of spin 1/2 particles, and they are showing remnant Fermi-Dirac behavior: no two identical fermions can ever be in the same state as defined by their four-momenta and spins no matter how they behave collectively.

## 173. Superfluidity

The odd number of constituents in He-3 (two protons, one neutron, and two electrons) classifies it as a fermion that obeys Fermi-Dirac statistics. So no two He-3 atoms can share the same quantum state defined by the four-momentum (energy and three-momentum) and spin. The surprise in the early 1970s was that He-3 can magnetically couple with another He-3 to form a boson and become a superfluid liquid at the extremely low temperature of

2.7 millikelvins. The He-3 pairs form one momentum macrostate. Because the component He-3 atoms are not bosons, there should be some small width to the macrostate momentum in addition to the small width because the He atoms are composed of fermions.

The pairs of atoms are magnetic, so the He-3 superfluid is more complex than its He-4 counterpart. In fact, superfluid He-3 exists in three different phases related to different magnetic or temperature conditions. In the A phase, for example, the superfluid is highly anisotropic—that is, directional like a liquid crystal.

*Scientific American. Special briefing on the Nobel Prizes in Science, "A New Superfluid."* Scientific American 276, no. 1 (1997): 15–16.

## 174. Gap Jumping

This Josephson effect is really quantum mechanical tunneling across the physical gap because the wave function for the superconducting pair extends beyond the end of the material into the gap and to the other side. If the superconducting material is actually in the form of a ring, then matching of the wave function for the pair around the ring must be made, restricting their angular momentum quantization to multiples of $h/2\pi$.

*Clarke, J. "SQUIDS."* Scientific American 271, no. 2 (1994): 46–53.

# 175. Nuclear Decay

The wave function extends through the potential barrier to the outside world. Therefore the probability to be outside the nucleus is not zero. So why does the wave function itself extend into the barrier? All confinement problems, classical and quantum, have solutions with functions that extend into the barrier, usually decreasing exponentially to almost zero within a few wavelengths. Atomic particles have relatively long wavelengths compared to the barrier thickness. So why does the wave function itself not end in the barrier? Because the effective barrier height decreases with radial distance.

The probability to tunnel through the barrier is proportional to Exp $[-Ar\sqrt{(U(r)-E)}]$, where $E$ is the energy of the incident particle, $U(r)$ is the barrier potential as a function of distance $r$, and A is a constant that includes Planck's constant h. Some closely related problems to be treated as tunneling through a barrier are:

1. Bare copper wire is cut and the two ends are twisted together. In spite of the fact that the copper is coated with copper oxide, the twisted ends still conduct electricity readily.

2. Tunnel diode operation.

3. Scanning tunneling microscope.

For a discussion about how a nuclear decay rate may be influenced by its environment, see the Peres reference below.

*Halliday, D., and R. Resnick.* Fundamentals of Physics. *New York: John Wiley & Sons,* 1988, pp. 1009–1010.

*Peres, A. "Zeno Paradox in Quantum Theory."* American Journal of Physics 48 (1980): 931–932.

# 176. Total Internal Reflection

Yes, the light goes a little beyond the interface. One can treat this behavior either classically or quantum mechanically. In QM the wave function for the photon extends beyond the glass-air interface into the air.

You can see this behavior in the following manner. Fill a drinking glass partially full with water. Tilt the glass and look down into it at the side wall at such an angle that the light entering your eye has been totally internally reflected from the wall. The wall will look silvery when this condition holds. Then press your moistened thumb against the outside of the glass. You will see the ridges of your fingerprint because, at those points, you will have interfered with the total reflection process. The valleys between the ridges are still far enough away from the glass that the reflection here remains total and you simply see a silvery whorl.

# 177. Annihilation

Fermi's Golden Rule hints that we should consider the phase space available for the final particles, and this phase space is related to the entropy of the final particles. If the entropy of the final state is greater than the entropy for the initial state, the process occurs. In the simpler case, when an electron at rest and its antiparticle, the positron at rest, annihilate each other, two photons are produced to conserve quantum numbers as well as energy and momentum. The entropy of the products is greater than the reactants. Why? Because there is much freedom in the direction of the photon polarizations. The interacting particle and antiparticle begin with their spins opposite but along a specific direction, thereby having a total spin of zero. In the final state with two identical photons emerging in opposite directions in order to conserve energy and linear momentum, the photon spins are opposite—that is, both spin +1 or both spin −1 with respect to their momentum directions—but the polarization vectors can be in any direction in the plane perpendicular to the momentum directions.

# 178. A Bouncing Ball

We present a simplified version of the complexities of this bouncing ball action. The compression of the ball (and the concrete being struck) sends phonons (quantum sound waves) running around telling the material that compression is occurring and that the increased energy density in parts of the ball can be reduced by expanding back to its normal size. Of course, the expansion overshoots and the ball "rings" as it leaves the concrete, each extended state also increasing the energy density in parts of the ball. One can model much of this behavior by assuming that the atoms and molecules are in a potential well somewhat similar to the parabolic well of the harmonic oscillator. However, instead of a potential energy for an atom versus the atomic separation distance being proportional to $r^2$ only, there must be additional terms proportional to $r^3$, etc., where $r$ is the distance from the equilibrium location.

Eventually the phonons help the ball get back into its normal shape, but the atoms and molecules never quite make their initial relative positions again, there being some residual distortion. Even the concrete being struck by the ball never quite recovers. Witness the eventual wear of a concrete highway by cars and trucks compressing the road, a more vigorous process but conceptually the same.

## 179. The EPR Paradox

There seems to be no classical thinking that would reproduce the data set. A predetermined instruction set would be akin to an algorithm for generating random numbers—but no such sets of numbers are truly random. One must accept the conclusion that Nature is quantum mechanical and therefore classical physics is only an approximation. The rules of QM agree with the results, but the details are too complicated mathematically to present herein. The references provide the extended discussion.

Even more surprising is the suggestion that locality is violated. That is, information from the first detector passes to the second detector without passing through imaginary spherical surfaces surrounding each, as if more dimensions exist in our world! Someone, someday, will determine a fundamental reason for this behavior of nature.

Eberly, J. H. "Bell Inequalities and Quantum Mechanics." American Journal of Physics 70 (2002): 276–279.

Einstein, A., B. Podolsky, and N. Rosen. "Can Quantum-Mechanical Description of Physical Reality Be Considered Complete?" Physical Review 47 (1935): 777–780.

Mermin, H. D. "Is the Moon There When Nobody Looks? Reality and the Quantum Theory." Physics Today 38 (1985): 38–47.

Shimony, A. "The Reality of the Quantum World." Scientific American 258, no. 1 (1988): 46–53.

Silverman, M. P. A Universe of Atoms, an Atom in the Universe. New York: Springer-Verlag, 2002, pp. 92–102.

von Baeyer, H. C. Taming the Atom: The Emergence of the Visible Microworld. Mineola, N.Y.: Dover, 1992, pp. 210–211.

## 180. Information and a Black Hole

For certain, one should worry about quantum information loss, especially if quantum mechanics is to provide a complete explanation for everything in the world. Does the black hole information increase with the inclusion of the chair? Let's see. A black hole has mass, spin, and possibly electric charge, weak charge, or color charge. That's all! We cannot determine the information content of the black hole from these quantities only. That is a problem. The most likely solution that would prevent quantum information loss is that the surrounding space just outside the event horizon of the black hole takes care of the information equation to make everything correct, emitting particles to compensate correctly.

The actual physics calculation of information change in the gravitational field of a black hole is much more complex and difficult. Among the necessary concerns is the fact that the black hole has performed a non-unitary transformation on the state of system when it devoured the chair. A non-unitary evolution is excluded in a quantum theory because it fails to preserve probability—that is, after a non-unitary evolution, the sum of the probabilities of all possible outcomes of an experiment may be greater or less than 1. Quantum mechanics could not survive. Perhaps the QM of a black hole will eventually be done and quantum gravity will save us from this catastrophe!

Bekenstein, J. D. *"Information in the Holographic Universe."* Scientific American 289, no. 2 (2003): 58–65.

# Chapter 9
# Can This Be Real?

## 181. Carbon-14 Dating

The ratio of C-14 to C-12 in living organisms will depend on many factors, including the local climate and the amounts of C-14 in the atmosphere, factors that can vary on time scales as short as tens of years. The radiocarbon dating process assumes in its zeroeth order approximation no variation in these factors over hundreds and thousands of years. But the cosmic ray intensity reaching the atmosphere may vary considerably, so the amount of C-14 produced will also vary. As the variations in the cosmic rays are determined by other independent methods, they can be incorporated into the C-14 dating as adjustments.

According to research literature, tree ring counts indicate that C-14 dating has fluctuations of the C-14 concentration in the atmosphere between 1400 and 1700 B.C.E. Furthermore, a comparison of radiocarbon-determined ages with ages of archaeological materials accurately established by other methods reveals that for the period from 100 B.C.E. to 1400, radiocarbon dating gives values that are too large, and that prior to 100 B.C.E. the radiocarbon values are too small.

At about 1600 B.C.E., the C-14 date values are about 175 years (5 percent) too small, increasing to about 300 years (6 percent) at 3000 B.C.E. The discrepancy appears to be a result of slight variations in Earth's magnetic field over the years, which would alter the cosmic ray intensities and hence C-14 production in the atmosphere. These corrections allow C-14 dates to be corrected, and even for 100,000 years ago the radiocarbon dates are good to within 5 percent.

*Staff of McGraw-Hill, eds.* McGraw-Hill Encyclopedia of Science & Technology. *Vol. 15. New York: McGraw-Hill, 2002, pp. 136–144.*

## 182. Nuclear Energy Levels

Even the shell model, often called the independent particle model of the nucleus, fails to correctly predict many of the energy level spacings unless the spin-orbit LS interactions are included. That is, the proton and neutron magnetic moments interact with magnetic fields produced by their orbital motions. These LS interactions add terms to the approximate constant potential of the shell model to dominate the quantum state sequence inside the nucleus. As a result, many energy levels change their relative positions on the energy scale, with levels from different principle quantum numbers becoming interchanged! Once the LS interaction was properly accounted for, all its predictions were shown to agree with the empirical data.

This model of the nucleus also explained why nuclei containing an even number of protons and neutrons are more stable than others. Like the energy levels for the electrons in quantum states outside the nucleus, the Fermi exclusion principle allows two identical particles per quantum state only. The nuclear quantum states for the protons are separate from the nuclear quantum states for the neutrons, and any particular state is filled when there are two identical particles with opposite spins. The proton levels are higher in energy than the corresponding neutron levels because there is the added Coulomb repulsion. Any extra proton or neutron can be added, but this additional particle must occupy a higher energy state, usually leading to an unstable nucleus.

Jolie, J. "Uncovering Supersymmetry." Scientific American 287, no. 1 (2002): 70–77.

Serway, R. A. Physics for Scientists & Engineers with Modern Physics, 3rd ed. Philadelphia: Saunders, 1990, pp. 1352–1354.

Tipler, P. A. Modern Physics. New York: Worth, 1978, pp. 427–432.

## 183. Nuclear Synthesis

Look at the binding energy curve for the elements and you will see that at least one isotope of Ni is well bound. Unfortunately, this isotope has a rapid decay mode. In fact, all the Ni isotopes from Ni-49 to Ni-57 have half-lives of only milliseconds to at most 10 days.

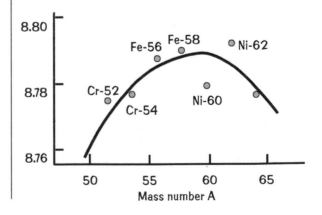

Although the championship of nuclear binding energy is often attributed to Fe-56, this isotope actually comes in third. The most tightly bound of the nuclei is Ni-62. The binding energies are 8.790 MeV/nucleon for Fe-56 and 8.795 MeV/nucleon for Ni-62. The binding-energy curve shows those nuclides that are close to the peak.

The most tightly bound nuclides are all even-even nuclei. Fe-56 is about a factor of ten more abundant in stars than Ni-62. The Fewell reference below indicates that the reason lies with the greater photodisintegration rate for Ni-62 in stellar interiors. Others have suggested that the very low rate of multistep production of Ni-62 from Co-59 is the culprit.

Fewell, M. P. "The Atomic Nuclide with the Highest Mean Binding Energy." American Journal of Physics 63 (1995): 653–658.

Shurtleff, R., and E. Derringh. "The Most Tightly Bound Nucleus." American Journal of Physics 57 (1989): 552.

# 184. Heavy Element Synthesis

The synthesis of the heavier elements beyond Fe is done during supernova explosions, in a few days or less, and the atomic debris are spewed out into space to later collect into new stars and planets and be there for incorporation into life forms.

The fusion process for elements up to Fe in the periodic table yields energy, and thus they occur in the normal stellar burning cycles. But since the "iron group"—those elements with isotope mass number of about A = 60—is at the peak of the binding energy curve, the fusion of elements above Fe *requires* energy, with the exception of the most tightly bound isotope—Ni-62, for example.

The elements beyond Fe are expected to be formed in the cataclysmic explosions known as supernovae in which a large flux of energetic neutrons build up mass approximately one unit at a time to produce the heavy nuclei. Following neutron capture, some isotopes beta decay to change a neutron into a proton plus an electron and an electron antineutrino, increasing the atomic number by one unit. Some sample sequences are:

Fe-56 + n → Fe-57 (stable)

Fe-57 + n → Fe-58 (stable)

Fe-58 + n → Fe-59 → Co-59 by beta decay

Co-59 + n → Co-60 → Ni-60 by beta decay

In principle this process could continue indefinitely, but the elements beyond uranium (Z = 92) are all radioactive.

The layers of the star containing the buildup of heavy elements may be blown off by the supernova explosion to provide the raw material of heavy elements within expanding hydrogen clouds that much later can condense to form new stars, planets, and the stuff of life.

*Krane, K. S. Modern Physics, 2nd ed. New York: John Wiley & Sons, 1995, pp. 290–291.*

# 185. Neutron Decay

The failure of the neutron in a nucleus to decay is a quantum mechanical effect. According to quantum mechanics, the rate of decay is dictated by Fermi's Golden Rule, which states that the rate is proportional to the probability of decay (i.e., the absolute value of the square of the matrix element connecting the initial and final states) times the density of final states. Because the free neutron decays to a proton plus electron plus electron antineutrino, we know that the probability for this beta decay process is not zero and that there are available final states for the three product particles. Energy conservation dictates that the total final state energy equals the total initial energy of the free neutron.

Inside a nucleus, the decay of a neutron is a transition from an initial energy state, the particular bound neutron state that the neutron occupies, to a final state consisting of a proton in some final proton energy state plus a free electron and a free electron antineutrino, the latter two particles contributing to the energy of the final state. Therefore, energy conservation dictates that the proton will be in a proton energy state that is lower in energy than the initial energy of the neutron. In many nuclei all available proton states—that is, those that are not occupied by protons—have higher energies than the energy of the initial neutron state, so the decay cannot occur.

The equivalent energy levels of the protons in nuclei are higher than for the neutrons because their energies include the Coulomb repulsion between two protons and other properties of the nuclear force, especially the spin dependence. Obviously the stable nuclei will include those for which neutron and proton decays do not occur!

*Asimov, I. Understanding Physics. New York: Hippocrene, 1988, p. 245.*

# 186. Finely Tuned Carbon?

The crucial energy comparison to make is not simply the radioactive state energy of 7.65 MeV to the practical limit value of 7.7 MeV, but one must include the comparison of the radioactive state energy 7.65 MeV to the energy 7.4 MeV of the reactants at rest. This energy of 0.25 MeV misses being too high for the production of carbon by the fractional amount of 0.05 MeV/0.25 MeV, or 20 percent, which is not so critical after all.

Barrow, J. D., and F. J. Tipler. The Anthropic Cosmological Principle. *Oxford: Oxford University Press, 1986, pp. 252–253.*

Livio, M., D. Hollowell, A. Weiss, and J. W. Truran. "The Anthropic Significance of the Existence of an Excited State of C-12." Nature 340, no. 6231 (1989): 281–284.

Weinberg, S. Facing Up: Science and Its Cultural Adversaries. *Cambridge, Mass.: Harvard University Press, 2001, pp. 235–237.*

# 187. Proton-Proton Cycle

The other stars are using the carbon cycle for their fusion energy. The common proton-proton cycle reaction is not the source of fusion energy in many stars burning hydrogen because the first reaction in this sequence has two protons combining to form a deuteron H-2, a very unlikely event that occurs slowly. A more likely sequence of reactions involves having C-12 be a catalyst:

$$C\text{-}12 + p \rightarrow N\text{-}13 + \gamma$$

$$N\text{-}13 \rightarrow C\text{-}13 + e^+ + \nu$$

$$C\text{-}13 + p \rightarrow N\text{-}14 + \gamma$$

$$N\text{-}14 + p \rightarrow O\text{-}15 + \gamma$$

$$O\text{-}15 \rightarrow N\text{-}15 + e^+ + \nu$$

$$N\text{-}15 + p \rightarrow C\text{-}12 + He\text{-}4$$

Called the carbon cycle, this sequence of reactions occurs much more rapidly than the proton-proton cycle sequence because the C-12 acts as a catalyst, neither being produced nor consumed by the totality of reactions. The net process is still the same: 4 protons $\rightarrow$ He-4, and the net energy produced is the same, but the rate of energy production is much higher.

The carbon cycle occurs at a higher temperature than the proton-proton cycle because the C and H Coulomb repulsion is greater than the H and H repulsion, so the Sun, with its internal temperature of about $15 \times 10^6$ K, is too cool to activate the carbon cycle, which requires about $20 \times 10^6$ K.

Krane, K. S. Modern Physics, *2nd ed. New York: John Wiley & Sons, 1995, pp. 282–285.*

# 188. Oklo Nuclear Reactor

The reaction sequence shows how to breed Pu from local U-238, which is the most abundant naturally occurring uranium isotope. Initially, neutrons come from the fission of U-235. However, the very high abundance of U-238 means that this isotope will absorb some of the neutrons to become U-239, decay by beta decay to neptunium 239, and then decay to Pu-239. Some of the resulting Pu-239 undergoes fission.

$$U\text{-}238 + n \rightarrow U\text{-}239 \rightarrow Np\text{-}239 + e^- + anti\text{-}\nu$$

$$Np\text{-}239 \rightarrow Pu\text{-}239 + e^- + anti\text{-}\nu$$

However, because the natural reactors at Oklo probably operated for such a long time, the Pu-239 had time to decay by alpha decay to U-235. Thus the Oklo natural reactors were true breeder reactors, fissioning more U-235 than originally existed in the reactors. The evidence for the breeder process remains in the reactor as more of the fission products than could possibly be produced by the amount of U-235 that has been lost from each of the reactor sites.

A second piece of evidence for Pu fission is the isotopic composition of the fission products in the mass range 100 to 110. To breed Pu and additional U-235, the reactors must have operated for periods significantly greater than the half-life of Pu 239, about 24,360 years.

*Cowan, G. A. "A Natural Fission Reactor." Scientific American, no. 7 (1976): 36–47.*

# 189. Human Radioactivity

At one time in the history of radiation safety, before extensive and long-term measurements, the recommended radiation limit was much less than the limit today. During those times, in the mid-1900s, two people in close proximity would have been emitting enough gamma radiation to exceed the recommended limit.

We can estimate the exposure amount and compare its value to the recommended limit today. There are approximately $10^5$ decays of K-40 isotopes per second in your body, but the decay chart tells us that only about 11 percent yield a gamma ray, producing about 1100 self-inflicted gamma rays per second, amounting to about 0.36 mSv per year, well below the recommended limit today. Even a group of 10 people closely packed would not provide a radiation exposure more than 3.6 mSv per year. So we are not radiation dangers to ourselves nor to our friends!

*Cohen, B. L. "Catalogue of Risks Extended and Updated." Health Physics 61 (1991): 317–335.*

# 190. Nuclear Surprises?

Both are true statements.

1. The only emissions from a nuclear power plant are (a) water vapor from its cooling towers, (b) thermal energy in the external cooling water, (c) any stray gamma rays not shielded (unlikely to be above normal background), (d) any radioactive isotopes created in the external cooling water (unlikely to be above normal background), and (e) electrical energy.

The emissions and safety procedures at a coal-burning power plant are not as strict and, because all coal naturally contains radioactive material with many isotopes, some of these radioactive isotopes escape into the air when the coal is piled in storage, when the coal is burned, and so on. Measurements at coal-burning plants verify that radioactive atoms and molecules are released.

Scientific researchers in the McBride et al. reference below have concluded from measurements "that Americans living near coal-fired power plants are exposed to higher radiation doses than those living near nuclear power plants that meet government regulations. . . . The fact that coal-fired power plants throughout the world are the major sources of radioactive materials released to the environment has several implications. It suggests that coal combustion is more hazardous to health than nuclear power and that it adds to the background radiation burden even more than does nuclear power. It also suggests that if radiation emissions from coal plants were regulated, their capital and operating costs would increase, making coal-fired power less economically competitive."

G. J. Aubrecht, in the reference below, states that the radioactivity danger from each coal-burning electrical plant is at least 100 times the danger from each nuclear plant.

2. Background radiation levels combining terrestrial (from K-40, Th-232, Ra-226, etc.) and cosmic radiation (photons, muons, etc.) are fairly constant over the world in the range of 8–15 μrads per hour. Assuming maximum damage to human tissue, this present background radiation level corresponds to about 1.8 mSv per year.

If one spreads all the human-produced artificial radioactive materials equally around the surface of Earth, the local increase in radioactivity is expected to be minuscule compared to this indigenous natural radioactive background. Suppose we had a million metric tonnes of human-made radioactive material to be dispersed over Earth of approximately $5 \times 10^{14}$ m$^2$. Each square meter would acquire an additional $0.2 \times 10^{-5}$ kg of radioactive material, compared to the natural amount of radioactive material in the top 10 centimeters of about $2 \times 10^{-2}$ kg, producing an insignificant amount of local radiation unless the half-lives were short, on the order of minutes to days. The additional amount adds only 1 part in 10,000 when dispersed around the globe.

Aubrecht, G. J. Energy, 2nd ed. Upper Saddle River, N.J.: Prentice Hall, 1994.

Cohen, B. L. "Catalogue of Risks Extended and Updated." Health Physics 61 (1991): 317–335.

Eisenbud, M. Environmental Radioactivity: From Natural, Industrial, and Military Sources, 4th ed. San Diego, Calif.: Academic Press, 1987.

McBride, J. P., R. E. Moore, J. P. Wither-spoon, and R. E. Blanco. "Radiological Impact of Airborne Effluents of Coal and Nuclear Plants." Science 202 (1978): 1045.

# 191. Cold Fusion

Cold fusion at room temperature is a real but unlikely possibility. The key idea is the quantum mechanical overlap of the wave functions of two nearby H-2 nuclei, for example. Their wave functions always overlap, no matter how far apart they are. However, the bigger the value of the wave function overlap, the more probable will be the possibility of the fusion process to make a He-4 nucleus.

Of course, there is a Coulomb barrier to be overcome. In the 1940s came the proposal that muonic atoms—a proton with a muon replacing the electron—might allow fusion because the muonic atom ground state puts the muon so close to the nucleus on average that the muonic atom appears neutral to the approaching proton. However, calculations have shown that the produced He isotope decays too quickly for this scheme to succeed in fusion energy production.

In gaseous form at room temperature, two colliding H-2 nuclei do not get close enough for a large wave function overlap, being strongly repelled by electrical forces acting between two positive nuclei. In a solid, however, at room temperature, H nuclei in neighboring lattice sites experience enormous accelerations, as large as $10^{14}$ m/s$^2$ in random directions. Sometimes these accelerations are toward each other, so the two protons can approach very close and perhaps fuse into a He nucleus. However, the actual calculation reveals the rarity of this event.

Despite the extreme improbability of deuteron fusion at room temperatures, so-called cold fusion, research groups worldwide continue its pursuit, as revealed in the references below.

Iwamura, Y., T. Itoh, M. Sakano, and S. Sakai. "Observations of Low-Energy Nuclear Reactions Induced by D$_2$ Gas Permeation through Pd Complexes." Infinite Energy 47 (January–February 2003): 14–18.

Mallove, E. F. "The Triumph of Alchemy: Professor John Bockris and the Transmutation Crisis at Texas A&M." Infinite Energy 32 (July–August 2000): 9–24.

Miles, M. H., B. F. Bush, and J. J. Lagowski. "Anomalous Effects Involving Excess Power, Radiation, and Helium Production during D$_2$O Electrolysis Using Palladium Cathodes." Fusion Technology 25 (1994): 478–486.

# 192. Fission of U-235

There are two major problems to be overcome in designing a fission device. The neutron distribution in a pure U-235 solid would decrease as the inverse distance squared from each nuclear decay source, and the target nuclei would be moving away during the expansion, so one has a diffusion problem complicated by moving targets. The moving targets contribute at least two difficulties: the density of targets is rapidly decreasing, and the neutron-capture cross section is a function of neutron kinetic energy as seen from the reference frame riding with each U-235 nucleus. Without the proper neutron capture rate by the receding U-235 nuclei, the chain reaction fizzles out.

Of course, the nuclear device cannot be expected to be pure U-235 because the isolation of enough quantities of U-235 from U-238 is too difficult and too costly. Therefore, there is mostly U-238 in the expanding solid with some U-235, so we have all the previously listed problems to solve but also must account for the nuclear properties of the U-238 as well as the U-235.

Apparently the Germans during World War II did not solve these diffusion problems satisfactorily.

Bernstein, J. "Heisenberg and the Critical Mass." American Journal of Physics 70, no. 9 (2002): 887–976.

# 193. Minimal Nuclear Device

Pure U-235 can be accumulated into a critical mass for a sustained nuclear chain reaction, but pure Pu-239 cannot start itself because the loss rate of neutrons exceeds the production rate. On the average, each U-235 fission produces 2.5 neutrons for every incident neutron. At the critical mass of fissile material the chain reaction will be sustained. For U-235 this critical mass is about 7 kilograms for ideal behavior, requiring a sphere about the diameter of a baseball of pure U-235. Surely this baseball would be too hot to handle!

Diffusion problems of an expanding material would require a neutron-reflecting strong tamper material surrounding the U-235 sphere to delay the expansion for a few microseconds to achieve additional fissions before exploding. At 100 percent efficiency the explosion would be equivalent to about 120 kilotonnes of TNT. However, no nuclear device is that efficient.

Declassified records indicate that about 60 kilograms of highly enriched uranium was used in the nuclear device that was released over Hiroshima, Japan, in 1945. The explosive charge for the device detonated over Nagasaki three days later was provided by about 8 kilograms of plutonium-239 (>90 percent Pu-239).

Bernstein, J. "Heisenberg and the Critical Mass." American Journal of Physics 70, no. 9 (2002): 887–976.

Pochin, E. Nuclear Radiation: Risks and Benefits. Oxford: Clarendon Press, 1983.

# 194. Large Nuclei

Small nuclei that become excited and deformed prefer to lose their energy by breaking up into helium nuclei (alpha particles) or C-12 nuclei whenever possible. In fact, researchers often talk about "nuclear molecules" composed of these two entities.

The larger nuclei, with more than 150 nucleons, usually spin faster when energy is added, and the result of a higher angular momentum state is a nucleus that is more deformed. As they de-excite, up to about 40 gamma rays are emitted by descending an "excitation ladder," producing a characteristic gamma ray emission spectrum. From this spectrum one can determine the nuclear angular momentum states and the nucleus's deformation shape. Superdeformed nuclei were discovered in this way.

Rotational motion of quantum objects such as atoms and molecules has a long and distinguished physics history. Quantized rotational motion of molecules was first recognized from the absorption spectra of infrared light in 1912. The occurrence of rotational motion of atomic nuclei first became a topic of interest in the late 1930s in an effort to explain observed nuclear excitation spectra by physicists Edward Teller and John Wheeler in about 1938.

Quantum mechanics dictates the shapes. Upon excitation, the nucleus first deforms into a shape like a rugby football, with a length-to-height ratio of about two to one. Mg-24 appears to behave as if two C-12 nuclei are its major components and seems to behave as a superdeformed nucleus in this rugby football shape. The next state would have an elongated hyperdeformed shape as a result of perhaps six alpha particles lined up along the long axis. This nucleus is highly unstable, and this nuclear sausage would produce an unmistakable debris pattern.

Recent detailed investigations of several Pb isotopes have yielded surprises. The angular distribution and polarization of the gamma rays show that they were not electric quadrupole (E2) transitions but magnetic dipole (M1). Classically, M1 radiation is pictured as being emitted from a rotating current loop, with the field oscillating at the same frequency as the frequency of rotation. Similar gamma-ray emission bands have recently been identified in other nuclei in the mass region around 110, where the nuclei also are nearly spherical. These spectra have a pattern that is typical of transitions between rotation states, which poses an awkward problem: how can we explain these regular patterns of M1 gamma rays? Apparently there is much more to understand.

Clark, R. M., et al. "Evidence for 'Magnetic Rotation' in Nuclei: Lifetimes of States in the M1 Bands of Pb-198, Pb-199." Physical Review Letters *78 (1997): 1868.*

*Macchiavelli, A. O., et al. "Semiclassical Description of the Shears Mechanism and the Role of Effective Interactions."* Physical Review C. *57 (1998): R1073.*

*Nolan, P. J., and P. J. Twin. "Superdeformed Shapes at High Angular Momentum."* Annual Review of Nuclear and Particle Science. *38 (1988): 533.*

## 195. Human Hearing

The Mössbauer Effect has been used to determine these actual displacements of an eardrum. The Mössbauer Effect utilizes the recoil-less emission of a 14.4 KeV gamma ray (photon) from an Fe-57 nucleus, say, and this gamma ray is normally absorbed by an Fe-57 nucleus in another object in its path. When the emitters (Fe-57 atoms placed on the eardrum) are moving with the eardrum, the emitted gamma rays pass through the second object, a cooled thin film of Fe containing some Fe-57 atoms, to be captured in a gamma-ray photon detector.

The important physical property here is that the natural linewidth of the emitted gamma ray is very narrow, about $10^{-8}$ eV, so that the Fe-57 recoil energy of about 0.002 eV produces a Doppler shift so large that no absorption in the cooled, thin film normally occurs. One can cancel this Doppler shift with a moving absorber or emitter of only 0.0002 m s$^{-1}$. Therefore, when the eardrum moves forward toward the stationary cooled thin film, there will be some resonance absorption of the gamma ray, so the detector count will decrease. When the eardrum moves opposite, there is no absorption. Because the eardrum vibrates in a nonlinear fashion, the details are somewhat more complicated. From the geometries and the physical properties of the emission and the Mössbauer absorption, the displacement values of the eardrum can be calculated. The sensitivity of this technique allows eardrum displacements that are only fractions of a nuclear diameter to be detected.

## 196. 1908 Siberia Meteorite

Willard Libby and Edward Teller explored this event with a reasonable hypothesis, since no rocky debris was ever found, and the amount of damage was enormous. If the meteorite were made of antimatter, then the ensuing matter-antimatter annihilation in the atmosphere and during the ground

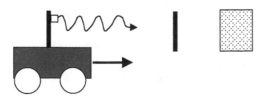

collision would create plenty of energetic photons at 0.511 MeV, 935 MeV, etc., due to electron-positron annihilations, proton-antiproton annihilations, etc. Many of these photons would interact with nitrogen N-14 in the atmosphere directly to make an excited N nuclear state or indirectly via secondary neutron production in the atmosphere. The excited state of N-14 decays to C-14, which increases the atmospheric concentration of C-14 in the carbon dioxide that is taken in by plants immediately after the event.

An increase in the C-14 to C-12 ratio should appear in the radiocarbon dating of living organisms such as plants, beginning in the year 1908 for local trees, and this increase in the ratio should appear a few years later for trees in North America, caused by atmospheric mixing of the C-14. One of us (F. P.) was working for the summer in Willard Libby's laboratory and was assigned to carefully separate the tree rings from an old oak tree, putting pieces into vials, and then coding the vials so that only I knew which vials contained which tree rings. The samples were radiocarbon-dated, and then the results were plotted by the C-14 to C-12 ratio versus the calendar year.

W. Libby, E. Teller, R. Berger, L. Wood, and F. Potter did not publish their radiocarbon-dating results, which demonstrated a significant increase in the 1911 C-14 to C-12 ratio detected in the old oak tree from Wisconsin. Later, a research group of C. Cowan, C. R. Atluri, and W. Libby (1965) did publish a similar result for the analysis of C-14 content in a 300-year-old Douglas fir from Arizona showing an increase in C-14 in 1911 with the same interpretation, supported by R. V. Gentry (1966). However, C-14 measurements of a tree by J. C. Lerman, W. G. Mook, and J. C. Vogel (1967) nearer the blast failed to show an increase in the 1909 ratio.

Several other interpretations of the 1908 meteorite event are possible. One of the main proposals is that an ice-rock comet struck Earth, much like comet Schumaker-Levy struck Jupiter in 1994. Also, one cannot rule out the possibility that a massive rocky meteorite just burned up completely—that is, broke into small fragments that burned up before striking the ground.

Chyba, C., P. Thomas, and K. Zahnle. "The 1908 Tunguska Explosion: Atmospheric Disruption of a Stony Asteroid." Nature 361 (1993): 40–44.

Cowan, C., C. R. Alturi, and W. F. Libby. "Possible Anti-matter Content of the Tunguska Meteor of 1908." Nature 206 (1965): 861–865.

Gentry, R. V. "Anti-matter Content of the Tunguska Meteor." Nature 211 (1966): 1071–1072.

Lerman, J. C., W. G. Mook, and J. C. Vogel. "Effect of the Tunguska Meteor and Sunspots on Radiocarbon in Tree Rings." Nature 216 (1967): 990–991.

## 197. The Standard Model

As far as we know, no such definitive argument for matching specific families exists in the Standard Model of Leptons and Quarks and their interactions. As long as six leptons cancel out the anomalies of the six quarks, for example, all is well! Indeed, one can use the second family of quarks to cancel the anomaly contributions from the first family of leptons, the third family of quarks to cancel the second family of leptons, and the first family of quarks to cancel the third family of leptons. In fact, any permutation of the traditional lineup of cancellations would succeed.

This ambiguity in the cancellations probably indicates that the Standard Model as understood is incomplete. One would expect the traditional scheme, but the conceptual understanding provided by the Standard Model does not dictate uniqueness.

One of us (F. P.) has proposed an interesting mathematical argument for matching lepton families to quark families based on correlations among finite rotational subgroups of the Standard Model gauge group for the leptons and quarks. In this scheme, each lepton family and each quark family is in a unique subgroup, and the one-to-one correlations are dictated by the

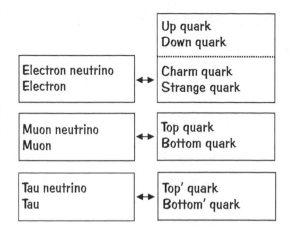

mathematics. Although the proposed scheme successfully predicted the mass of the top quark, this geometrical basis for the Standard Model awaits confirmation of other specific predictions for collisions, which are under way at Fermilab and soon to be done using the Large Hadron Collider.

Glashow, S. L. "Quarks with Color and Flavor." Scientific American 233, no. 4 (1975): 38–50.

Kane, G. "The Dawn of Physics beyond the Standard Model." Scientific American 288, no. 6 (2003): 68–75.

Liss, T. M., and P. L. Tipton. "The Discovery of the Top Quark." Scientific American 277, no. 3 (1997): 54–59.

Potter, F. "Geometrical Basis for the Standard Model." International Journal of Theoretical Physics 33 (1994): 279–306.

## 198. Spontaneous Symmetry Breaking

Yes. At least two other methods can achieve the same symmetry-breaking result without requiring the Higgs particle. The Standard Model is described by its continuous gauge group $SU(3)_C \times SU(2)_W \times U(1)_Y$. The simplest way of all is to spontaneously break this continuous group to a discrete symmetry subgroup of the continuous group $SU(2)$. That is, the lepton and quark flavor eigenstates would be associated with finite rotational subgroups of $SU(2)$ instead of the continuous group. An analogy from geometry would be to begin with a sphere and then symmetry-break to a regular tetrahedron, or a regular octahedron, or a regular icosahedron. Reconciling discreteness with the continuous symmetry group $U(1)$ of quantum electrodynamics may be a problem, however, where phases are assumed to vary continuously. Another symmetry-breaking approach is the quark condensate method, which also does not require a Higgs particle.

At present, no Higgs particle has been detected at the accelerators, even though its mass is expected to be below 200 GeV/c$^2$, within the energy range of the large accelerators. Of course, the decay of such a Higgs particle is a flavor-changing neutral current weak decay, which means that its decay rate is severely suppressed, so only a few Higgs decays would have been detected among the particle debris so far. When the Large Hadron Collider comes online in 2005 or later there should be copious production of the Higgs

particle if this mechanism is truly the source of symmetry breaking and the particle masses. If the Higgs particle does not show up, then spontaneous symmetry breaking to a discrete group remains an alternative possibility.

Coleman, S. Aspects of Symmetry. Cambridge, Eng.: Cambridge University Press, 1985, pp. 113–130.

Icke, V. The Force of Symmetry. New York: Cambridge University Press, 1999, pp. 232–248.

Potter, F. "Geometrical Basis for the Standard Model." International Journal of Theoretical Physics 33 (1994): 279–306.

't Hooft, G. "Gauge Theories of the Forces between Elementary Particles." Scientific American 242, no. 6 (1980): 104–140.

## 199. Proton Mass

Quantum chromodynamics describes the interactions of the quarks. The up and down quark masses are listed as ~ 5 MeV/c$^2$ each. However, these "current" quarks are not what is meant by having them confined inside a proton by the color fields. Instead one must use the effective mass—the "constituent" mass—which accounts for this confinement and which can be estimated from the Heisenberg uncertainty principle. Since $\delta x \delta p_x \geq h/4\pi$, and each quark is confined within the proton radius of about one Fermi, we estimate $\delta p_x \sim 100$ MeV. In three dimensions, the total $dp \sim \sqrt{(\delta p_x)^2 + (\delta p_y)^2 + (\delta p_z)^2} \sim 170$ MeV/c$^2$. So at least 510 MeV/c$^2$ of the proton mass is to be associated with the "constituent" mass of the three quarks within the proton. The remainder is the energy contributions of the gluons holding the proton together.

Most of the properties of protons, *except the spin,* seem to be determined by these three "valence quarks," much like the valence electrons determine the important chemical properties of atoms. However, when the proton's innards are probed more energetically, more structure is found, up to four or five more particles, called "virtual quarks." In addition, up to 30 gluons can be detected. The proton is revealing its inner sanctum to investigators, and the view is becoming quite interesting. Quarks, antiquarks, and gluons can be said to form a thick "soup" inside the proton, and theoretical and experimental physicists are working together to figure out the recipe.

Today we know that the three valence quarks cannot alone account for the proton's spin. The whole "sea" of quarks, antiquarks, and gluons each possess spin, so one must first determine the contribution made by each individual member of this seething mass. The results so far suggest that the sea of quarks makes a minimal contribution to the overall spin of a nucleon!

Abbott, D., et al. "Measurement of Tensor Polarization in Elastic Electron-Deuteron Scattering at Large Momentum Transfer." Physical Review Letters 84 (2000): 5053.

Aniol, K. A., et al. "Measurement of the Neutral Weak Form Factors of the Proton." Physical Review Letters 82 (1999): 1096.

# 200. Right- and Left-handed Neutrinos?

No. The weak interaction is associated with the SU(2)-weak part of the Standard Model gauge group that operates in the unitary plane—a plane with two complex axes. That is, particle fundamental lepton and quark states are defined in this unitary plane. All rotations in the normal unitary plane involve only left-handed doublets and right-handed singlets, dictated solely by the mathematics of the geometrical transformation. Mathematicians call these transformations right and left screw operations. So the physical property of left-handed doublet states for the weak interaction is dictated by the mathematical property of rotations in the unitary plane. Nature simply "knows" the mathematics!

The antiparticle eigenstates are in the conjugate unitary plane, which is gauge-equivalent (*not* equivalent) to the normal unitary plane, so the energy values of particles and antiparticles are the same, but all other properties are opposites. In this conjugate unitary plane the mathematics dictates right-handed doublets and left-handed singlets. The existence of two gauge-equivalent but different 2-D complex spaces conjugate to one another dictates that the universe has both particles and antiparticles. Why there exist so many more particles than antiparticles in our present universe remains to be resolved.

*Altmann, S. L.* Rotations, Quaternions, and Double Groups. *Oxford: Clarendon Press, 1986, pp. 121–123.*

# 201. Physics without Equations

The best way to use cellular automata (CA) on computers is to incorporate the fundamental interactions of the Standard Model of Leptons and Quarks plus the gravitational interaction of the general theory of relativity, or preferably its quantum gravitational version when available. We know that all these fundamental interactions in nature correspond mathematically to *local* phase changes, a process that can be simulated with CA without using equations by using a clever enactment of the path-integral approach to doing all of physics in real time. Not yet fully achieved except by very crude approximation, the physics of many-particle interactions will be accomplished by large-scale grid computation methods or perhaps by the equivalent on a quantum computer.

The fundamental idea is to determine the present behavior of a particle by summing over all the phase information from its local environment. Of course, each particle also provides phase information to its environment both near and far. The particle's new location is the region where the phases match best. The calculation game requires a dynamic limit to how many nearby cells are counted in order to accumulate a good approximation of the phase information and to maintain the local geometrical symmetry. A proof-of-concept calculation has been done by one of us (F. P.) on a desktop computer using thousands of nodes in a 3-D array, but a good calculation requires millions of cells or the equivalent.

The marriage between physics and mathematics has been a happy and fruitful one over many centuries. Mathematical equations, from simple algebraic ones to the more challenging differential equations, have allowed us to summarize an enormous amount of physical phenomena into a simple format. The underlying fundamental symmetries of nature have been the true source of many of these equations. However, formulating these symmetries as the Schrödinger equation and Maxwell's equations, for example, and solving the equations are human processes. We cannot expect Nature to do the same when the simpler process of looking locally for information is more direct. Therefore we think that understanding the universe by combining CA with path integrals will be the physics of future generations.

*Icke, V.* The Force of Symmetry. *New York: Cambridge University Press, 1999, pp. 178–206.*

# Chapter 10
# Over My Head

## 202. Olbers' Paradox

German astronomer Heinrich Olbers (1758–1840) was not the first scientist to ask "Why is the night sky dark?" but his name remains connected to this paradox. The night sky is dark because the time required for the radiation field to reach thermodynamic equilibrium is large compared to all other time scales of interest—that is, the lifetime of stars is far too short for the sky to be as bright as the paradox suggests. In addition, if all the matter in the universe were converted to radiation, the equilibrium temperature of the universe would be about 20 K, illustrating that there is insufficient energy to have a bright sky. Edward R. Harrison in the early 1970s determined this solution and also determined that the usual explanation, based on a cosmological redshift of the light from distant sources, was not needed even though its argument would likewise produce a dark night sky.

The critical quantity is the ratio of the average lifetime $t_{ave}$ of a star to the time $T$ required for the universe to reach thermodynamic equilibrium. Starting with a uniform density of stars, an observer can appreciate that after a clock time $t = t_{ave}$ there will be an expanding sphere of burned-out stars beyond which lies a shell of luminous stars. The radiation from this shell has a maximum radiation density equal to the surface radiation density from the average star times the ratio $t_{ave}/T$ as long as the clock time $t << T$. But $t_{ave}$ is at most a little more than 10 billion years, while $T$ can be shown to be tens of billions of years, so the night sky remains dark. Harrison shows that this argument is true for all present models of the universe and does not require a cosmological redshift. He argues that Lord Kelvin (1901) was the first to give the correct answer, which Edgar Allan Poe had anticipated in his qualitative cosmological speculations. For the detailed calculations, see the references below.

*Harrison, E. R. "Why the Sky Is Dark at Night."* Physics Today 27, no. 2 (1974): 30–36.

———. *"The Dark Night-Sky Riddle."* Science 226 (1984): 941–945.

*Pesic, P. "Brightness at Night."* American Journal of Physics 66 (1998): 1013–1015.

## 203. Headlight Effect

In the special theory of relativity (STR) a result called the headlight effect occurs. One considers the Lorentz-Fitzgerald contraction of distances in the direction parallel to the constant velocity and no change in the perpendicular direction. If the primed frame is the vehicle frame, then the angle in the two frames are related by $\cos \phi = (\cos \phi' + v/c)/(1 + v/c \cos \phi')$. Substituting the appropriate values tells us that $\cos \phi \sim 1$, or $\phi \sim 0°$! Therefore, all the light is in a very small solid angle in the forward direction, and only an observer directly along the line of motion will see the light. You will not see the light from the relativistic vehicle passing nearby unless your eye is within the very narrow light cone.

In the rest frame of the source, the star emits light in all directions, yet the calculation reveals that for an observer of a very fast-approaching star, practically all its light will be shining along the direction of motion! A fast-approaching star or galaxy will possess a very narrow bright headlight beam that could miss Earth. Meanwhile, a fast-receding star or galaxy may not be seen at all because its light is redshifted out of the visible range and practically all its light shines away from us!

So in observing stars, there is this STR headlight effect to consider. There also are the different clock rates for the two frames of reference, so the number of photons emitted per second on the star and received at Earth will differ. Moreover, the spectrum of light will be different as well.

*Taylor, E. F., and J. A. Wheeler.* Spacetime Physics. *San Francisco: W. H. Freeman, 1966, p. 69.*

## 204. Incommunicado?

No and yes! No, in the normal sense case because the relative velocity can never exceed the speed of light. The successive pulses may arrive less and less often, but you will never outrun the light.

And yes, you would lose communication contact if we allow the space itself to expand, analogous to the expansion of the universe in present cosmological models. The addition of velocities is the old classical physics one, not the relativistic one. The photon velocity is affected by the local environment; the local substratum (i.e., coordinate system) "drags" the photon along. Imagine two local regions in rapid recession from each other. If the person in one region fires a photon toward the other, the substratum of the first region drags the photon along, slowing the photon's progress toward its target. If the expansion rate is high enough, the two regions can be receding from each other at light speed or greater, preventing communication between you and your friend.

Higbie, J. "Radial Photon Paths in a Cosmic Model: A Student Exercise." American Journal of Physics 51 (1983): 1102–1107.

# 205. Local Accelerations

The presence of the massive body can be determined by the trajectories of the two test masses upon their release. In the simple case in which the laboratory is not moving with respect to the massive body, when released equidistant from the object but separated from each other, the two test masses will move toward each other faster than their mutual gravitational acceleration as they fall toward the body. In addition, if they are separated vertically so that one test mass begins closer to the massive body than the other, their vertical separation distance will change as they fall. In a uniform gravitational field their separation distance would remain fixed in value in each test.

One can extend this problem to consider a rotating massive body. Can observers inside a spaceship determine by "local" measurements only—that is, without looking outside—if they are in the field of a rotating central mass, or if they are just moving with velocity V on a Schwarzschild background metric? Yes they can; by using at least four test particles inside their spaceship and having the capability to measure their relative accelerations, they can succeed in determining all the components of the Riemann tensor and decide whether they are in the space-time curvature of a rotating central mass. Note that gyroscopes do not help here because one would need to check their alignment with stars outside, which is forbidden. The challenge here is to measure a new effect, called intrinsic gravitomagnetism, introduced by the GTR, that the space-time geometry and the corresponding curvature invariants are affected and determined by *both* mass-energy and mass-energy currents relative to other mass—that is, by mass-energy currents that cannot be eliminated by a Lorentz transformation. See the Ciufolini and Wheeler reference below for the details.

Ciufolini, I., and J. A. Wheeler, Gravitation and Inertia. Princeton, N.J.: Princeton University Press, 1995, pp. 358–360.

Kalotas, T. M., A. R. Lee., and R. B. Miller. "Einstein on Safari." The Physics Teacher 29 (1991): 122–124.

Martin, J. L. General Relativity: A Guide to Its Consequences for Gravity and Cosmology. Chichester, Eng.: John Wiley & Sons, Ellis Horwood, 1988, pp. 93–94.

# 206. Twin Paradox

Both special theory of relativity (STR) and general theory of relativity (GTR) explanations for the aging of the space-traveling twin should be considered. If by the STR we consider inertial reference frames only and ignore any accelerations experienced by the space traveler, a symmetry would exist between the two frames, and the twins must both age at the same rate. Therefore, the accelerations experienced by the space traveler make the difference in the aging.

One can handle these accelerations in the STR or in the GTR. Some people argue that this twin paradox problem requires only the STR because there is no curved space-time in the problem—that is, both twins can be considered to be in a flat space-time because no gravitational accelerations near a mass are necessary. Then one would handle the accelerations for the space-traveler twin in terms of STR calculations, perhaps via the velocity parameter technique. A true GTR problem, by contrast, would require the physics of the curved metrics near a massive body.

The solution of the twin paradox using the GTR relies on clocks ticking slower in a gravitational potential near a mass. The clock at a far distance from the mass ticks at its fastest rate and, if brought closer to the mass, begins to tick slower and slower. Therefore a person closer to the massive body, where the gravitational acceleration is greater, ages slower.

In cases where the acceleration of a spaceship can be approximated by an equivalent gravitational acceleration— that is, using the Equivalence Principle—we can expect the traveling clock to tick slower during the acceleration. And that behavior is why the traveling twin ages less. As seen by a third observer at rest with respect to the stars and the stay-at-home twin, the clock on the spaceship is changing its rate of ticking during the accelerations.

## 207. Twin Watches

A watch ticks at its fastest rate when at rest and when there is no gravitational field. So there are two effects to consider: (1) from the special theory of relativity (STR), the motion of the watch with respect to the laboratory frame affects the ticking rate; and (2) the change in gravitational potential according to the general theory of relativity (GTR) affects the clock ticking rate. For a watch in free fall, the two effects are exactly opposite and cancel! The two watches agree again when she takes the second reading.

Now for the details. First, are there any symmetry considerations that would simplify the calculation? Yes; the two parts of the journey for the moving watch—the upward and the downward parts—are time reflections of each other, and these two parts require the same elapsed time in the laboratory frame and in the moving-watch frame.

Pick the laboratory frame of reference. As the watch goes upward in the lab frame at its maximum velocity initially, the STR makes the watch tick faster as the velocity decreases, and the GTR makes the watch tick faster as greater height is achieved. On the downward journey, the watch ticks slower and slower by both STR and GTR effects. So we need only calculate the changes in the tick rate when the watch has gone upward by a small amount—$\Delta h$, say.

From the STR, the time interval $T$ between ticks at velocity $v$ is given by $T = T'/\sqrt{(1 - v^2/c^2)}$, where $T'$ is the time interval between ticks of the watch in its own reference frame. At two different heights—$h_1$ and $h_2 = h_1 + \Delta h$—the time intervals between ticks are $T_1 = T'/\sqrt{(1 - v_1^2/c^2)}$ and $T_2 = T'/\sqrt{(1 - v_2^2/c^2)}$, respectively, because the velocities will be different at the two heights. Since $v \ll c$, and assuming a uniform acceleration approximation for free fall, by the third golden rule of kinematics, $v_2^2 = v_1^2 - 2 g \Delta h$. Substitute the velocities squared into the watch's time interval relations and expand the square roots in the denominators by the Taylor series expansion $1/\sqrt{(1 - \varepsilon)} \sim 1 + \varepsilon/2 + \dots$ . One calculates $T_2 \sim T_1 - T' g \Delta h/c^2$, a quantity proportional to the change in height.

From the GTR, the time interval $T$ between clock ticks at radial distance $R$ outside of a body of mass $M$ is given by $T = T' \sqrt{(1 - 2GM/(Rc^2))}$. In the limit of very large $R$, the clock ticks at its fastest rate. By definition, $g = GM/R^2$ at the surface of Earth. Substitute the above heights for the two distances from the massive body and take the difference. One calculates that $T_2 \sim T_1 - T' g \Delta h/c^2$, a quantity proportional to the change in height and a quantity from the GTR that changes as fast as the quantity from the STR.

So the total change in the tick rate going upward is canceled by the total change in the tick rate coming downward, to make no net change when they are once again at the same height. If this argument has any flaws, do not blame either of my colleagues, Richard P. Feynman (deceased) or B. Winstein, for they know not what they had wrought!

# 208. Global Positioning Satellites

The general theory of relativity (GTR) plays an important role! Corrections must be made for clock rates in a gravitational field in addition to the special theory of relativity (STR) corrections to the clocks for the movement of the satellite. Both relativistic effects foul up what should have been a pretty simple geometry calculation relating distance to time and velocity. The clocks in the satellites tick at a slightly faster rate than identical clocks on the ground because they are in a slightly weaker gravitational field, being farther from the center of Earth. They tick slower than the Earth-bound clocks because they are moving faster with respect to the stars.

We can estimate the sizes of these effects. From the STR, the time interval $T$ between the ticks of a clock moving at velocity $v$ is given by $T = T'/\sqrt{(1 - v^2/c^2)}$, where $T'$ is the time interval between ticks of the clock in its own reference frame. At slow speeds $v \ll c$, one expands the square root by the Taylor series expansion $1/\sqrt{(1 - \varepsilon)} \sim 1 + \varepsilon/2 + \ldots$ to obtain $T \sim T'(1 + v^2/(2c^2))$. For satellites orbiting Earth in about 720 minutes, their speed makes the time correction factor about $1.1 \times 10^{-10}$. Multiplied by the speed of light, this time factor corresponds to a distance error of about 3.3 centimeters for each second.

From the GTR, the time interval $T$ between clock ticks at radial distance $R$ outside of a body of mass M is given by $T = T'\sqrt{(1 - 2GM/(Rc^2))}$. Consider two different radii: $r_1 = 6.37 \times 10^6$ m and $r_2 = 2.02 \times 10^7$ m. Substitute the two radii for the two distances from the massive body and take the difference. One calculates a clock correction factor of about $4.8 \times 10^{-10}$ for this GTR effect, a little more than four times the STR effect, or about 14.4 centimeters of error every second. So in 10 minutes even this small effect produces an error of about 86 meters if not accounted for. Who would have thought that both STR and GTR effects are big enough to play important roles in such a useful practical system as GPS!

# 209. Solar Redshift

Even though there may be no relative radial motion between the Sun and the observer on Earth, there is still a gravitational redshift dictated by the general theory of relativity (GTR). Recall that the infinitesimal distance $ds$ in a flat Euclidean space with coordinates $(r, \theta, \phi)$ is defined by $ds^2 = c^2\, dt^2 - dr^2 - r^2\, d\theta^2 - r^2 \sin^2\theta\, d\phi^2$. In the gravitational field of mass $M$, this infinitesimal distance in the GTR becomes the Schwarzschild line element $ds^2 = (1 - r_g/r)\, c^2\, dt^2 - (1 - r_g/r)^{-1}\, dr^2 - r^2\, d\theta^2 - r^2 \sin^2\theta\, d\phi^2$, where $r_g = 2GM/c^2$ and G is the gravitational constant.

We see that near a massive body such as the Sun, the time coordinate includes a factor $\sqrt{(1 - rg/r)}$, where $r$ is the position of the light measured from the center of the Sun. By evaluating this factor at the Sun's surface and at Earth's distance, one finds that the clocks at the two distances are ticking at different rates, faster for bigger $r$. One approach is to assume that the photon does not change its inherent physical properties—for example, it maintains its characteristic frequency established during the emission process at the surface of the Sun. Then the observer on Earth, who has the faster-ticking reference clock with respect to the stars, will measure a lower photon frequency and see the light as redshifted.

A second approach determines that the change in $r$ for the photon can be shown to correspond to a change in gravitational potential. The photon essentially begins in a gravitational potential energy valley and climbs upward to reach Earth. Its total energy must remain constant, so the increase in gravitational potential energy is matched by the decrease in photon energy—that is, a redshifted photon—because its energy $E = h\nu$.

Krane, K. S. Modern Physics. *New York: John Wiley & Sons, 1983, pp. 438–442.*

# 210. Orbiting Bodies

The general theory of relativity (GTR) in the Schwarzschild metric approximation for the space-time metric about the Sun predicts a precession of the planetary orbit. Mercury, for example, accumulates a total GTR precession of about 43 seconds of arc per Earth century. There are many other precessional effects acting on the orbit, including effects from all the other planets orbiting the Sun, all these perturbations amounting to a whopping 532 seconds of arc per century, all but the residual 43 seconds being explained with Newtonian mechanics.

When the angular change around the orbit is calculated with GTR in the $\phi$-coordinate and then independently in the $r$-coordinate, there is disagreement, which is the conceptual source of the effect. Or one can assign an additional equivalent mass distribution for the energy in the gravitational field surrounding the Sun, creating a metric that does not correspond to a $1/r$ potential, perhaps $1/r^2$ or $1/r^3$, or some other function of $r$ instead of the Newtonian inverse $r$ potential. All these functions of $r$ will exhibit a precession of the orbit.

In addition, a body in orbit such as a planet orbiting the Sun actually does not obey Kepler's third law precisely. That is, even when we ignore the precession of the orbit by assuming its return to the same angle with respect to the stars, the period of orbit needs correction. This period of orbit correction is a so-called fourth independent general test of the GTR, in addition to the gravitational redshift, the deflection of starlight, and the precession of an orbit.

The correction to the period of orbit, as calculated by Preston and Weber in the reference below, begins by putting the reference clock at the center of the orbit. The classical Newtonian period of orbit is given by Kepler's third law: $T = 2\pi\, a^{3/2}/\sqrt{(GM)}$, where $a$ is the semimajor axis of the ellipse and $M$ is the mass value of the central body. For the elliptical orbit of eccentricity $\varepsilon$, in the GTR one can calculate the period of orbit in the radial coordinate $T_r = T\,(1/\alpha + 3/2\, r_g/r)$ and in the $\phi$-coordinate $T_\phi = T\,(1/\alpha - 3/2\,(r_g/r)\,(\varepsilon^2/\alpha))$, where $\alpha = (1 - \varepsilon^2)^{3/2}$. For orbiting bodies near potential black holes, this correction can get large as the radial distance $r$ approaches the Schwarzschild radius $r_g = 2GM/c^2$.

Krane, K. S. Modern Physics. *New York: John Wiley & Sons, 1983, pp. 438–442.*

Landau, L. D., and E. M. Lifschitz. The Classical Theory of Fields, *4th ed. Sydney: Butterworth-Heinemann, 1987, pp. 328–330.*

Preston, H. G., and J. Weber, "The Period of Orbit as a Test of General Relativity." Physics Essays 6 (1993): 465.

# 211. Gravitational Lensing

The general theory of relativity (GTR) tells us that all forms of energy are affected by a gravitational field, including the energy carried by photons of light. In the Schwarzschild metric surrounding a star, for example, the path of the light from a distant star is diverted toward the Sun from a straight-line path when passing near a massive body such as our Sun. Half of the deflection angle is caused by the Newtonian attraction of the Sun; the second half is caused by the geometrical modification, called curvature, of space by the Sun. Gravitational lensing ideas have been around for about 200 years, but only in the past decade or so has gravitational lensing played an important part in astronomical measurements.

If the intermediate massive body is a galaxy, then light from distance sources behind this galaxy will be focused somewhat by the two effects,

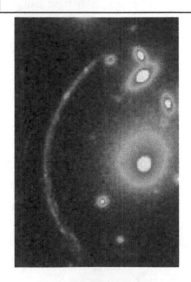

Hubble space telescope lensing image

just like the light going through a glass lens is refracted to a focus. However, the geometry of the light focusing is much more complicated for gravitational lensing than for a simple symmetrical convex lens, for several reasons. The light may be focused onto a line instead of a point, for example, in an ideal case. Therefore, astronomers can use galaxies as lenses to gather more light from far objects better. The focus may be poor, but the greater intensity allows many spectroscopic techniques to work better.

Usually the image resolution of the distant object is quite limited by inhomogeneities in the intermediate galaxy. However, these properties of the intermediate galaxy can be examined quite well! In fact, if the classical application of the GTR to calculate the focusing effects in gravitational lensing is correct, then the total mass of the intermediate galaxy and its mass distribution can be determined. Galaxy masses determined with gravitational lensing have disagreed with the very successful proposal for a modified Newtonian dynamics (MOND) within a galaxy region and have supported the "dark matter" models. However, the game is not over for MOND because large-scale gravitational quantization, either from a version of M-theory and superstrings or from some other quantization scheme, may come to the rescue eventually.

Wambsganss, J. "Gravity's Kaleidoscope." Scientific American 285, no. 5 (2001): 64–71.

# 212. Cosmological Redshifts

There are three distinct causes for the spectral shift of light emitted (or absorbed) by a galaxy: the kinematical Doppler shift of the special theory of relativity (STR), the gravitational redshift of the general theory of relativity (GTR), and the cosmological redshift caused by the expansion of the universe. These three effects cannot be distinguished from one another by observing the spectrum of a single galaxy or other single light source. One can separate out the kinematical Doppler shift for a cluster of galaxies, however, via statistical methods.

The standard explanation of the cosmological redshift says that the coordinate system of the universe is expanding while the galaxies remain at their local coordinate values. One can use an expanding balloon to "mimic" this type of behavior. Inflate a balloon enough to enable you to draw a coordinate system on its surface. Place some galaxies on the balloon surface. Now inflate the balloon further. The galaxies are farther apart, but they maintain the same coordinate values. Or one can use Earth. If Earth began to expand, Philadelphia and Los

Angeles would move apart, yet each would retain its present longitude and latitude. Distances are stretched, so wavelengths will become longer.

Perhaps another viewpoint will be helpful. One does not need to expand space in this view. According to S. Weinberg in the Chown reference below, simply accept the fact that "every bit of the universe is rushing away from every other bit"—that is, "the galaxies are exploding away from each other, as any cloud of particles would do if they are set in motion away from each other." The matter inside the individual galaxies does not take part in the general expansion because the local gravity holds the local matter together. The universe's expansion appears just beyond the frontiers of the Local Group of galaxies, about 4 million light-years from the Local Group's center of mass.

M. L. Bedran in the reference listed below compares the Doppler redshift to the cosmological redshift for a galaxy with $z = 1$, where $1 + z = \exp [v/c]$, determining that a 2.4 percent difference between the two redshifts exists for this galaxy with an STR value of $v/c = 0.6$.

Bedran, M. L. "A Comparison between the Doppler and Cosmological Redshifts." American Journal of Physics 70 (2002): 406–408.

Chown, M. "All You Ever Wanted to Know about the Big Bang." New Scientist (April 17, 1993): 32–33.

# 213. Tired-Light Hypothesis

The only two specific pieces of evidence are the time dilation arising from the expansion of the universe, and the spectral shape of the cosmic microwave background. Astronomers see that exploding stars in distant galaxies brighten and fade more slowly than those nearby. If the star emits a light pulse on January 1 and a second pulse on February 1, these two pulses are separated by one light-month. As they travel toward Earth, their separation distance increases, perhaps doubling, so that they are received two months apart. The tired-light hypothesis cannot explain this extended time interval. In fact, distant supernovas are observed to wax and wane more slowly than nearby ones.

The observed spectrum of the microwave background radiation is a perfect blackbody shape, easily explained by the expansion of the universe from a thermodynamic equilibrium condition. For the tired-light hypothesis, an initial blackbody spectrum does not remain a blackbody spectrum as the light becomes redshifted.

Croswell, K. The Universe at Midnight: Observations Illuminating the Cosmos. New York: Free Press, 2001, p. 76.

# 214. Black Hole Entropy

There should be radiation from the black hole—that is, from the surrounding space, not from inside the black hole, because nothing can get out. This Hawking radiation was first calculated in the 1970s and awaits experimental verification.

By taking quantum mechanics into account, particles and antiparticles are being created continually via virtual pair creation in the vacuum. When this process occurs near a black hole, one particle of the pair may be "eaten" by the black hole and the other may escape. In the thermal equilibrium state, the amount of energy that the black hole loses to Hawking radiation is exactly balanced by the energy gained by swallowing other "thermal particles" that happen to be running around in the "thermal bath" in which the black hole finds itself.

The temperature of a nonrotating black hole is given by $T = hc^3/(8\pi kGM)$, where h is Planck's constant and k is Boltzmann's constant. Note that this expression connects gravitation, thermodynamics, and quantum mechanics. For black holes of a few solar masses, the temperature is only about $10^{-6}$ K! The smaller black holes with little mass will be at a much higher temperature, contrary to intuition.

Hawking, S. W. A Brief History of Time: From the Big Bang to Black Holes. *New York: Bantam Books, 1988, pp. 104–110.*

Penrose, R., The Emperor's New Mind. *Oxford: Oxford University Press, 1989, pp. 361–363.*

# 215. Black Hole Collision

Yes. The two black holes should coalesce like two liquid drops. We need to ensure that the entropy afterward in the coalesced final state is greater than the entropy in the initial state. We do that by adding up the entropy in the two states, separate black holes versus one larger black hole with gravitational waves carrying away some energy and entropy in the final state.

The black hole entropy is proportional to the event horizon area, which grows as the mass to the fourth power. Suppose we take two black holes, one of mass $M_1$ and the other of mass $M_2$. Their initial total entropy is proportional to $M_1^4 + M_2^4$. If they merge and their final total mass is approximately $M_1 + M_2$, then their final total entropy is proportional to $(M_1 + M_2)^4$, which you can verify is greater than the original total entropy, so the reaction will go. In cases where the final entropy is only slightly greater than the initial entropy for the black hole parts only, one may need to add in the entropy in the gravitational waves to ensure a bigger inequality

Thus a bigger mass means a bigger surface area, which means a bigger entropy than if the two smaller black holes remained apart. Simulations in 3-D of colliding black holes and their emitted gravitational waves can be seen on the Internet.

Bekenstein, J. D. "Black-Hole Thermodynamics." Physics Today 33, no. 1 (1980): 24–31.

# 216. Centrifugal Force Paradox

The conceptual resolution of this paradox starts with the consideration of light paths near the black hole. The general theory of relativity (GTR) predicts that there should be light paths around the black hole that are circular at a radial distance of 1.5 times the gravitational radius $r_g$ = $2GM/c^2$. Around one of these circular light paths imagine a circular tube centered exactly on the path of the circular ray

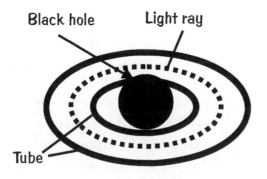

so that the axis of the tube and the path of the ray coincide. Measurements with a straight ruler verify that the axis of the tube is circular, yet because of the bending of the light rays, the tube is seen as absolutely straight by an observer on the axis. A lamp held at the axis by a colleague will appear dimmer to you as he or she walks away along the axis, but the lamp is never obscured, so you must conclude that the tube is straight. Therefore, along this circular path one would expect no centrifugal force effects.

Instead of the tube being around the circular light path, consider the tube to be around a smaller circular path centered on the black hole. With rulers one can again verify that the tube curves to the left, with the black hole on the left as one walks forward. The outward direction is to the right. Everyday experience tells us that the centrifugal force pushes outward. Again your colleague walks away with the lamp held along the axis of the tube. If somehow the light rays were not bent by the gravitational field of the black hole, you would see the lamp disappear behind the left side of the tube, and you would conclude that the tube bends to the left. If the path is the one discussed above, the lamp is always in view. But the tube is so close to the black hole that the light rays bend even more than circular rays. So you actually see the lamp disappear to the right. Therefore the outward direction is to the left, and you must predict that the centrifugal force would push to the left!

Several related paradoxes for paths near a massive body are discussed in the references, including the fact that for a rocket to maintain a constant speed the boosters would need to fire perpendicularly to the path, but their force is not dependent on how fast the rocket is moving.

Abramowicz, M. A. "Black Holes and the Centrifugal Force Paradox." Scientific American 268, no. 3 (1993): 74–81.

Abramowicz, M. A., and J. P. Lasota. "On Traveling Round without Feeling It and Uncurving Curves." American Journal of Physics 54 (1986): 936–939.

## 217. Geodesics and Light Rays

The two statements are not in conflict. One must always distinguish geodesics in four-dimensional space-time from geodesics in three-dimensional space. Light rays always follow geodesics in 4-D space-time, but these paths are not necessarily geodesics in 3-D space. An analogy is helpful. Each great circle on a globe is a geodesic line on the two-dimensional surface but, being a circle, the great circle is not a geodesic line in the 3-D Euclidean space in which the globe sits.

In conventional geometry, the geodesic is the shortest curve between two points measured by counting how many rulers fit along the curve. In a flat space—that is, in a space free from gravitational fields—the geodesic is a straight line. In the GTR one can define the distance between two points in space as half the time it takes for light to travel from one point to the other and back, multiplied by the speed of light. In flat space, the two definitions agree.

In 4-D space-time, light always moves along geodesics and traces the geometry of space-time. In a 3-D space warped by a gravitational field, however, the light rays are curved and do not coincide with geodesics in general, so the geometry of space is not traced by light rays.

Abramowicz, M. A. "Black Holes and the Centrifugal Force Paradox." Scientific American 268, no. 3 (1993):74–81.

Abramowicz, M. A., and J. P. Lasota. "On Traveling Round without Feeling It and Uncurving Curves." American Journal of Physics 54 (1986): 936–939.

Misner, C. W., K. S. Thorne, and J. A. Wheeler. Gravitation. San Francisco: W. H. Freeman, 1973, pp. 31–34.

## 218. Galaxy Rotation

To retain a Newtonian gravitation explanation for the rotation properties of a galaxy, additional matter, called "dark matter," in a halo surrounding the galaxy has been proposed, its mass being about 10 times the mass of the visible mass! Known particles such as electrons, protons, neutrinos and so on cannot be its major constituents; otherwise the particles in the halo would have been detected already. Some exotic form of matter/energy is required in this galactic halo. However, any type or types of possible "dark matter" in sufficient quantities has not yet been found.

An interesting proposal that does not require "dark matter" is called MOND, an acronym for modified Newtonian dynamics, which applies whenever the inward radial acceleration in the galaxy is below the value $a_0$ = $-1.1 \times 10^{-10}$ *m s*$^{-2}$, an incredibly small amount by Earth standards but a value that occurs in most galaxies. Essentially, MOND replaces the Newtonian acceleration $g_N$ by $g = \sqrt{(g_N a_0)}$ . All galaxies examined so far seem to obey the consequences of this ad hoc rule by using the galactic visible matter only, but its possible origin in terms of fundamental physics principles is being investigated still. The major problem for MOND has been its inability to accommodate the empirical results on the focusing of distant starlight by gravitational lensing.

An even more exotic solution has been proposed to explain galaxy rotation without requiring "dark matter."

Briefly, the large-scale structure and behavior of the galaxy may result from the galaxy being in some quantization state. In modeling this type of theory, all the disk stars would be in the same quantization state independent of position radially and, by the virial theorem, must have the same tangential velocity $V = GM^2/(nJ)$, where $M$ is the amount of visible mass, $n$ is a small integer, and $J$ is the total angular momentum of this visible mass of the galaxy. Substituting reasonable values for our Galaxy (the Milky Way), for example, one obtains a value near to the measured value of $V$ = 220 km s$^{-1}$. This theory predicts that the next quantization state would have exactly half the disk tangential velocity and, indeed, in 2003 a mass current of stars circulating the Galaxy just beyond the "edge of the easily visible disk" with a tangential velocity of 110 km s$^{-1}$ was determined serendipitously from data collected by the Sloan Digital Sky Survey (SDSS)! Whether this proposed large-scale quantization faithfully represents gravitational behavior in galaxies and in the universe remains to be fully examined.

Cline, D. B. "The Search for Dark Matter." Scientific American 288, no. 3 (2003): 50–59.

Milgrom, M. "Does Dark Matter Really Exist?" Scientific American 287, no. 2 (2002): 42–52.

Preston, H. G., and F. Potter. "Exploring Large-Scale Gravitational Quantization *without h-bar in Planetary Systems, Galaxies, and the Universe."* E-print archive for physics papers. *http://lanl.arxiv.org/abs/gr-qc/0303112 (2003).*

# 219. Cosmic Background Radiation

Amazingly, cosmic background radiation (CBR) has a perfect blackbody distribution! This CBR is uniform and isotropic in the universe to one part in 100,000 and amazingly flat over large spatial regions—that is, large solid angles. One suspects that these large spatial regions, even those in opposite directions in the sky, have always been in communication with each other, to make them so uniform. Of course, smaller regions have their individual characteristic galaxies, clusters of galaxies, and so on.

The most popular interpretation, the standard inflationary model of the universe, requires the universe to originate with the far regions much closer to each other in thermal equilibrium for the observed uniformity to develop, then for a very fast inflation to occur that separated them out of communication reach. Now we see these galaxies, once close together, in opposite directions in the universe and large regions with the same large-scale characteristics in all directions. Their original collective blackbody spectrum at a high temperature now exhibits a low-temperature blackbody spectrum because the expansion of the universe has "stretched the wavelengths."

The stars we see are not in thermal equilibrium as a collective whole. One cannot produce a perfect blackbody spectrum at any temperature by simply using billions of stars that are not in thermal equilibrium as a whole and adding up their radiation intensities in the universe. One cannot obtain a blackbody spectrum from many other hypotheses about the cosmological redshift of light from distant objects, such as the tired-light effect wherein the light loses energy during tranversal of the universe.

One could, however, speculate that the galaxies have never changed their average separations, that there has been no coordinate expansion as stated in the standard inflationary model. The cosmological redshifts would correspond to the redshift produced by an "effective cosmological gravitational potential well"—for example, in which the source sits lower in the well than the observer, true for all sources and all observers in the universe. The galaxies would be closer together at all epochs, in communication with each other, and in thermal equilibrium, so the measured uniformity would be expected—that is, all directions should look the same.

One consequence would be that one could never see galaxies beyond about 12 billion light-years or so, the actual distance value depending on the average matter/energy density of the vacuum. The redshifts would be interpreted as "effective recession velocities" that reach light speed at this far distance.

Hasinger, G., and R. Gilli. "The Cosmic Reality Check." Scientific American 286, no. 3 (2002): 60–67.

Peebles, P. J. E. "Making Sense of Modern Cosmology." Scientific American 284, no. 1 (2001): 54–55.

# 220. Planetary Spacings

The orbital radii of the planets only roughly follow the Titius-Bode law, so this specific pattern is probably bogus. However, L. Nottale and his research group have shown that the planets obey a generalized Schrödinger-like wave equation (with one unknown parameter) that has solutions dictating a regular pattern for where orbiting bodies reach an equilibrium radius. The planets of the Solar System occupy only these radial positions and leave some equilibrium radii unoccupied, perhaps a consequence of their history of formation. However, even though Nottale's fits are extremely good, there are several other sets of small integers that statistically fit as well as the set proposed by Nottale, including many sets with larger integers.

Extrasolar planetary systems with three planets have been found, but their statistical fits allow several sets of integers also, so they are not yet the definitive test. Unfortunately, we must wait for a definite extrasolar system or a precise laboratory test to resolve the issue of whether the patterns are simply numerology or an exhibit of part of a new fundamental gravitational theory.

Lynch, P. "On the Significance of the Titius-Bode Law for the Distribution of the Planets." Monthly Notice of the Royal Astronomical Society 341 (2003): 1174–1178.

Nottale, L. "Scale-Relativity and Quantization of Extra-Solar Planetary Systems." Astronomy & Astrophysics 315 (1996): L9–L12.

Nottale, L., G. Schumacher, and J. Gay. "Scale Relativity and Quantization of the Solar System." Astronomy & Astrophysics 322 (1997): 1018–1022.

# 221. Entropy in the Big Bang

We quote from the reference listed. The "standard" answer attempting to explain the paradox is:

True, the fireball was effectively in thermal equilibrium at the beginning, but the universe at that time was very tiny. The fireball represented the state of maximum entropy that could be permitted for a universe of *that* tiny size, but the entropy so permitted would have been minute by comparison with that which is allowed for a universe of the size that we find it to be today. As the universe expanded, the permitted maximum entropy increased with the universe's size, but the actual entropy in the universe lagged well behind this permitted maximum. The second law arises because the actual entropy is always striving to catch up with this permitted maximum.

This answer cannot be correct if the universe will eventually suffer a "big crunch," for then the argument would apply again in the reverse direction! We are at an impasse.

*Penrose, R. The Emperor's New Mind. Oxford: Oxford University Press, 1989, pp. 317–330.*

# 222. Gravitational Wave Detectors

For all kinds of waves, for locations beyond several wavelengths, the solutions of the wave equation correspond to the radiation field transporting energy and momentum from the source into the surrounding space. When considering possible sources of gravitational waves in the Galaxy and beyond, the wavelengths are typically several kilometers or more. One could place the rotating laboratory gravitational wave source several kilometers away or more from the gravitational wave detector, but the decrease in the radiation field intensity with distance squared combined with the low sensitivity of the detectors make this arrangement unlikely to work with present detectors. Therefore, as far as we know, there has never been a true test of the gravitational wave response of any detector to gravitational radiation using laboratory sources of gravity waves.

There have been two fundamental types of gravitational wave detectors: the Weber bar antenna, named after pioneering physicist Joseph Weber, who began this research field in the 1950s with his one meter diameter suspended aluminum bar; and the interferometer type such as LIGO, first analyzed by the same Joseph Weber and his students. The classical calculation of the resonant response of the Weber bar reveals just how limited is its sensitivity to gravitational waves originating in our Solar System and Galaxy. However, if the Weber bar antenna actually behaves differently than originally expected, as a collective quantum oscillator responding coherently, say, then it will respond well to all frequencies of the incident gravitational waves. Hundreds to thousands of vibrational modes could be excited in a large range of frequencies with an increase in sensitivity of many powers of ten.

At this time there has been no substantiated detection of gravitational waves by either type of detector. Weber reported a twice-daily response of his two side-by-side almost-identical bar antennas for orientations pointing toward the center of the Galaxy during a period of almost two decades, but no other researcher has verified this behavior with an independent detector. So we must wait for the first detection of gravitational waves by LIGO or other detectors. Unfortunately, interferometer types of detectors such as LIGO and VIRGO cannot operate as a collective quantum oscillator.

## 223. Space Curvature

The proposed method for determining the curvature of space will work for both continuous and discrete spaces. If we assume a uniform density of stars, or galaxies if we choose to count galaxies, the number $N$ of this particular kind of source within a sphere of radius $R$ in a Euclidean space (zero curvature space) is given by $N = \rho \, 4\pi R^3/3$, where $\rho$ is the uniform density. When $N$ is plotted against the distance, $N$ will fall short, match, or

exceed the cubic curve for the three types of spaces: positively curved, flat, or negatively curved, respectively.

On a "small" scale, when the total number of sources is less than a few hundred, there can be a relatively large uncertainty in the general behavior of the plotted curve. But as more and more sources are counted at farther distances, the asymptotic behavior should become apparent. However, adjustments must be made for the finite velocity of light and for possible evolutionary changes in the sources. Distant sources are sampled at an earlier time, possibly at a closer distance.

If the universe is actually representative of a discrete space, one can show that by counting many sources one can determine the curvature in the limit as the number of sources becomes large. Think of a lattice of points as one simple example, such as a regular lattice of atoms in a solid. By counting the nearest neighbors only, then the next nearest neighbors, and so on, one eventually approaches asymptotically to a plotted line from which the curvature can be determined.

Of course, in a discrete space, one must be careful not to count over and over the images of the same source. For example, imagine space divided into identical cubes next to each other and filling all space. Standing inside one cube, we can look to our right to see ourselves inside the first cube on our right looking away to the next cube, and so on. Each successive image will be dimmer and will be earlier in time because the light does not travel infinitely fast. If our real space in the universe is discrete, the cube size would be enormous, certainly way beyond the size of our Local Group of galaxies; otherwise we would have detected this discreteness already by having seen multiple images of our own Galaxy.

If space in the universe is curved, then cubes will not fill the space. Mathematicians point out that one of the dodecahedral spaces would be the simplest space-filling for a negatively curved discrete space, the most likely type of space curvature for the universe. However, the curvature of the universe is not known unambiguously yet, although a flat space with no curvature will nicely fit the present data in the standard model of an inflationary big bang universe.

Eckroth, C. A. "Counting Distant Radio Sources to Determine the Overall Curvature of Space." The Physics Teacher 30 (1992): 92–93.

Gruber, R. P., A. D. Gruber, R. Hamilton, and S. M. Matthews. "Space Curvature and the 'Heavy Banana Paradox.'" The Physics Teacher 29 (1991): 147–149.

Levin, J. How the Universe Got Its Spots: Diary of a Finite Time in a Finite Space. Princeton, N.J.: Princeton University Press, 2003, pp. 132–155.

Wolfram, S. A New Kind of Science. Champaign, Ill.: Wolfram Media, 2002, pp. 433–540.

# 224. The Total Energy

Yes, there can be the creation of matter out of nothing with no violation of conservation laws! First proposed in 1958 by H. Margenau and later recalculated in more detail by N. Rosen and others in 1994, the gravitational energy cancels the mass energy in a closed, homogeneous universe.

The simplest general approach was done by Margenau. Consider a finite spherical universe of radius $R$ filled with matter and radiation of equivalent total mass $M$. The gravitational potential energy is the negative quantity $-kGM^2/R$, where G is the gravitational constant and k is a positive numerical factor not greatly different from 1. The total energy $E$ in the universe is then $E = Mc^2 - kGM^2/R$. Using representative values such as $R = 1.3 \times 10^{26}$ m and a mass density of $8 \times 10^{-27}$ kg/m$^3$, we estimate k ~ 2.4 when $E = 0$. Nathan Rosen and others showed that the gravitational energy cancels out the mass energy without resorting to numerical estimates.

Cooperstock, F. I., and M. Israelit. "The Energy of the Universe." Foundations of Physics 25 (1995): 631–635.

Jammer, M. Einstein and Religion: Physics and Theology. Princeton, N.J.: Princeton University Press, 1999, pp. 201–203.

*Margenau, H.* Thomas and the Physics of 1958: A Confrontation. *Aquinas Lecture 23. Milwaukee: Marquette University Press, 1958, p. 41.*

*Rosen, N. "The Energy of the Universe."* General Relativity and Gravitation 26 (1994): 319–321.

# 225. Different Universes?

If the lepton and quark masses are dictated by fundamental mathematical quantities, we are behooved to consider that all fundamental quantities in nature have an origin in fundamental mathematics. There can be no alternative universes, each supposedly having different fundamental constants, for they must have the same mathematics dictating the same physical constants.

In 1994 F. Potter, within the confines of the Standard Model of Leptons and Quarks, related the lepton and quark mass ratios to a mathematical invariant called the elliptic modular invariant J, which is invariant under all linear transformations. The critical prediction is a fourth quark family with a b′ quark mass of about 80 $GeV/c^2$ and a t′ quark mass of about 2,600 $GeV/c^2$. Although searches for a b′ quark have been under way at the Fermilab collider for several years, its existence cannot be ruled out yet because the decay reactions have very low probability and will be overwhelmed by many other particle decays into the same final products. Perhaps when the Large Hadron Collider is turned on in a few years, with its very high rate of production of quarks, the statistics will be so much better that the b′ quark should be easy to find.

If the b′ quark is found, then we expect that all other fundamental physical constants should also be derivable from mathematical invariants. If the proposed scheme is correct, then our universe is the only universe possible. Even exotic speculations such as time travel could be eliminated if the direction of time is one of the innate properties of the particle state definition. However, we must remember always that Nature is more clever than we can hope to be, so we must continue to test every reasonable proposal for the truth.

*Linde, A. "The Self-Reproducing Inflationary Universe."* Scientific American 279, no. 11 (1994): 48–55.

*Potter, F. "Geometrical Basis for the Standard Model."* International Journal of Theoretical Physics 33 (1994): 279–306.

*Tegmark, M. "Parallel Universes."* Scientific American 288, no. 5 (2003): 40–51.

# Chapter 11
# Crystal Blue
# Persuasion

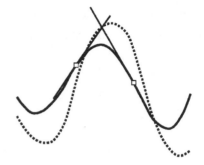

## 226. Iodine Prophylaxis

The iodine tablets, usually potassium iodine, work by "topping up" the thyroid gland with stable iodine to reduce its accumulation of any radioactive iodine that may have been released into the environment by a nuclear accident. Inhalation of the radioactive iodine in the air is the major route of entry into the body and the thyroid. For maximum benefit, the iodine tablets should be taken before the radioactivity fallout reaches your area; otherwise the tablets themselves will have become radioactive.

## 227. Bicycle Tracks

In "The Adventure of the Priory School," Sherlock Holmes not only draws a map of the neighborhood of the school but also examines several sets of tyre tracks on the moor. He needed to determine the bicycle's direction of travel solely from his inspection of the tracks. Holmes could tell from the depth of the wheel impression which track was made by the rear tyre. You don't have that information, but a little mathematics reveals the answer.

Note that the rear wheel of a bike always points to the place where the front wheel touches the ground. Therefore, the tangent to the rear wheel track will always cross the front wheel track, while the front wheel track does not exhibit this geometrical property.

Once we identify the rear track, we can pick two random points on it and extend the tangents to where they cross the front track in both directions, measure the segments, and determine which direction yields segments of the same length. Since a bicycle cannot change its length, we learn the direction of travel.

## 228. Earth Warming

Yes, there can be fluctuations in the thermal energy conducted and convected to Earth's surface from sources within. There are several types of thermal energy sources, including radioactive nuclei emitting particles that transfer their kinetic energy to thermal energy, as well as friction between internal rock flows, which could create local hot spots and/or temporary changes in the flow properties of liquid rock or the thermal conductivity of the rocks. Therefore, small changes in the rates of thermal energy transport to Earth's surface are possible and most likely happen continually. Are these internal sources the culprits in the present slow average temperature increase? These fluctuations are expected to be too small, but no one knows for sure.

## 229. Frequency Jamming

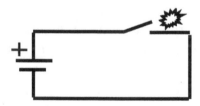

A spark gap usually is a wonderful noisy source of electromagnetic waves at all frequencies simultaneously. The greater the current across the gap, the more intense will be the total radiation emitted at each frequency. There will be a distribution of intensity versus frequency that can be "tuned" a little by adjusting the gap spacing.

A simple spark gap would be a small battery, such as a D cell, and two

wires in the process of making contact. Held near a radio, the small spark across the gap can be heard through the radio, indicating that many frequencies are being emitted. As an additional demonstration, one can even move a radio near a small electric motor of the kind that has brushes to hear its rotation frequency because the brushes make and break contact each revolution.

Of course, if one desires to have a higher-current spark gap, a car battery or a transformer can be used with proper safety precautions to provide a healthy current that can be operated in an intermittent mode or in a continuous mode. Nearby radios, televisions, and so on will be affected by this larger-current spark gap device. Even GPS transmissions between 1,000 MHz and 2,000 MHz may be affected, so some care must be taken not to violate federal transmission limitations.

## 230. Light Energy

A light source emitting light of frequency $f$ approaching an observer at constant velocity $v$ will appear blue-shifted by an amount corresponding to the relativistic Doppler effect formula $f' = f \sqrt{[(1-v^2/c^2)]/(1-v/c)}$ because the clock ticking rates will be different for the source reference frame and the observer reference frame, *and* their separation distance is decreasing.

When v << c, we can expand the formula in a Taylor series to obtain $f' \sim f(1 + v/c - v^2/2c^2 + \ldots)$, so the leading term in powers of v/c is positive, corresponding to the blueshift. We assume that an acceleration itself does not produce an additional fundamental frequency shift, although there will be acceleration effects because the source is changing instantaneous comoving inertial frames.

For a photon, its energy is $E = hf$ and its momentum is $p = E/c$, so both energy and momentum are different in different reference frames because the observed frequencies are different. Notice that the recoil of the source on emission of light and of the observer on detection are not accounted for in the discussion and that the energy and momentum input necessary to keep the relative velocity of the source and observer fixed must be considered also. Of course, energy and momentum conservation laws are obeyed in this example.

# 231. Acid Rain

Not so! The falling raindrops will not remain neutral at pH 7. Pure rainwater falling through unpolluted air is an acid, with a pH of about 5.6, because as the drops form and fall they dissolve carbon dioxide in the air and react to produce carbonic acid, $H_2CO_3$. Officially, therefore, acid rain is defined to have a pH of less than 5.0, a condition that tends to occur more often in industrialized areas of the world than in remote regions.

Human activities can increase the amount of $CO_2$ in the air, but so do many natural resources, such as volcanoes, lightning strikes, cows, bacteria, and fires. When industrial and automobile exhausts release sulfur compounds and nitrogen compounds, these molecules combine with oxygen to form sulfuric acid and nitric acid, which can harm ecosystems, historic monuments and buildings, and the health of people around the world. The reduction of sulfur and nitrogen compounds released into the air has become a worldwide concern.

*Trefil, J. The Nature of Science. Boston: Houghton Mifflin, 2003, pp. 6–7.*

# 232. Electrical Current

The electrons in house wiring move at a snail's pace, with an average drift velocity of about a millimeter per second. These electrons, which are free to move in the metal wires, are distributed throughout, so when the switch is closed to make a complete circuit, they move en masse, sort of like water passing through a continuous hose that closes on itself. The electron velocity is limited because its negative electric charge interacts with the lattice of positive ions during the movement.

In addition to having a drift velocity, the electrons experience a random sequence of pinball-like collisions to change their speeds and directions, essentially behaving as a free electron gas. Consequently, the host metal gains some thermal energy, and its temperature rises. If a lamp is the incandescent type, the tungsten alloy filament will gain enough energy by this process to dramatically increase its temperature to a new equilibrium temperature of about 2000 K to glow in the visible and the infrared.

# 233. Earth's Orbit

Although the general theory of relativity dictates a precession of the perihelion for all planets, including Earth, this effect is very small compared to perturbations provided by gravitational influences of all the planets. Apparently Earth's elliptical orbit will pass through a repetitive cycling about every 93,000 years from its present ellipse and orientation with respect to the stars, to a circle, then back to an ellipse with an orientation perpendicular to the present orientation, to a circle again, back to an ellipse perpendicular again, etc., until the present elliptical orientation is recovered approximately. Of course, all the planets are experiencing these perturbation effects simultaneously, so the detailed calculations become quite interesting!

# 234. Crystal Growth

The speed and precision of the crystal growth depend on many factors, including the temperature, concentration, and purity of the solution. Assuming the ideal solution, each additional atom to be added from the solution must first find a location on the growing surface of the developing crystal. But these atoms in solution are randomly moving about, making random collisions with the crystal at random locations on the surface. How can they build a perfect single crystal?

Their little secret is that some atoms that have been added at marginal locations, say, can escape from these surface locations to allow other atoms from the solution to find a better location nearby, "better" here meaning to be held electrostatically tighter to the crystal. But these better locations do not occur in chronological order because they are determined by the collective influence of numerous atoms already in the crystal, and the best position one microsecond ago may not be the best position for an atom now. Therefore, the addition and subtraction of atoms from the growing crystal surface proceeds almost by trial and error! Consequently, one cannot write down an algorithm for placing atoms from the solution onto the growing crystal.

When the crystal grows slowly,

there is plenty of time for the sampling process to proceed to fruition, and the crystals tend to form with fewer dislocations and inclusions. When the crystal grows rapidly, errors in the crystal structure become trapped, and these crystals tend to have many dislocations and inclusions.

# 235. Ruby, Sapphire, and Emerald

How are ruby, sapphire, and emerald crystals related? Ruby and sapphire are color variations of the same mineral, corundum. Rubies contain a small amount of chromium. Pure corundum is a colorless, trigonal crystal that occurs in a wide variety of colors due to infiltrations of other elements. All color variations of corundum, with the exception of ruby, are called sapphires.

Rubies are red variations of the mineral corundum, a crystalline form of aluminum oxide and one of the most durable minerals that exists. Only diamonds are harder. Rubies' rich, red colors arise from the substitution of a small number of aluminum atoms by chromium atoms. When exposed to high temperatures, rubies turn green, but they regain their original color after cooling. Some rubies phosphoresce with a vivid red glow when illuminated by ultraviolet light.

Sapphire is aluminum oxide with trace impurities of iron and titanium atoms, which are responsible for the deep blue color shades most people associate with sapphire. Several other colors of corundum, such as yellow, reddish-orange, and violet, also are classified as sapphire. Synthetic sapphires have been produced commercially since 1902 and are used for scratch-resistant watch crystals, optical scanners, and in applications where physical strength and transparency to ultraviolet irradiation are important.

Emeralds are quite different from rubies and sapphires in that emeralds are the green form of beryl, colored by the presence of chromium or vanadium. The crystal structure of beryl emeralds is hexagonal (six-sided), with a hardness slightly higher than quartz but considerably less than diamond. Emeralds are notorious for containing flaws, and flawless stones are rare and greatly valued.

The colors in these gemstones are produced by characteristic downward atomic transitions involving F-centers (from the German word *farbe,* meaning color) at the chromium or other atoms. In a simple view of an F-center, the ambient light excites one electron in the chromium atom, for example, so the atom can be treated analogous to the hydrogen atom, with a large average radius for the location of the electron away from the chromium nucleus. The electron transition back to a lower energy state causes the emission of a photon in the visible.

*Nassau, K. "The Causes of Color."* Scientific American 243, no. 4 (1980): 124–154.

*Perkowitz, S. "True Colors: Why Things Look the Way They Do."* The Sciences (May–June 1991): 22–28.

## 236. Kordylewski Clouds

Joseph L. Lagrange in the late 1700s calculated via Newton's laws that there are five special positions for objects bound by any two-body system. Now called Lagrange points, positions L1, L2, and L3 are unstable, while L4 and L5 are stable. Several spacecraft have been placed at or near these Lagrange points, and there have been proposals for building space colonies at the L4 or L5 positions.

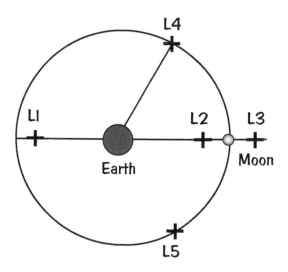

Applying Kepler's third law for a particle of mass $\mu$ between Earth's mass $m$ and the Sun's mass $M$ orbiting with Earth's period $T$, one obtains after several steps $GM\mu/(r{-}R)^3 - Gm\mu/(R^2(r{-}R)) = GM\mu/r^3$ where $r$ is the Earth–Sun distance and $R$ is the Earth–particle distance. The particle at L1 will be about 0.01 times the distance to the Sun. The L3 point on the night side of Earth can be calculated in the same way, replacing r–R with r+R. However, the other three points must be calculated with the gravitational attractions of the other planets included.

Similar calculations have been done for the five Lagrange points for the Earth-Moon system. Polish astronomer K. Kordylewski in 1961 reported the observation of dust clouds at the L5 point, but some observers have not seen them. Particles here may not remain long before being ejected, according to calculations.

*Kordylewski, K. "Photographische Untersuchungen des Librationspunktes Lim5 System Erde-Mond."* Acta Astronomica 11 (1961): 165–169.

*O'Neill, G. K. "The Colonization of Space."* Physics Today 27, no. 9 (1974): 32.

## 237. Twist Scooter

If the plane of the V arms of the twist bike remained horizontal at all times, there would be no forward motion except by pushing with a foot on the ground. By tilting the vertical shaft of the handlebar sideways about 10 degrees or so, the front of the scooter is lowered a bit, and a slight unbalanced outward push of the arms of the V by the rider's legs provides a forward net force similar to the resulting net force exercised by an ice skater. The initial forward movement from rest may be difficult on an upslope exceeding a particular angle determined by the size of the wheels and by the possible tip angle of the handlebar.

# 238. Unruh Radiation

The Equivalence Principle tells us that a particle accelerating in the vacuum is equivalent to a particle at rest in a uniform gravitational field. If there is radiation in one case, there must be radiation in the other equivalent case. The former is called Bekenstein radiation, the latter Unruh radiation, named after physicists who studied the properties of the radiation theoretically. No one has ever measured this radiation because its intensity is many powers of 10 too faint to be detected.

# 239. Star Diameters

One can determine the interference diameter of a distant star even when optical parallax resolution of its diameter is impossible by utilizing quantum interference between the photons from the left side of the star arriving out of phase with the photons from the right side. In other words, the photons are not expected to be in phase. Their phase difference depends on three parameters: their initial phase difference, the distance to the star, and the diameter of the star. By slowly changing the separation of the two photodetectors on the arms of an intensity interferometer, one can sweep across a range of phase differences to determine the diameter of the source. One laboratory analogy might be considering how one would determine the spacing between the slits of a two-slit interference experiment with a similar apparatus. Ultimately, the amplitudes and not really the intensities interfere. However, the phase correlations depend on the product of intensities, in contrast to the two-slit interference example.

The original experiment is known as the Brown-Twiss experiment, named after the two researchers who first succeeded in using the technique to determine a star diameter back in 1957. Interference associated with the superposition of separate light intensities was viewed with considerable skepticism. Apparently, as the story goes, one of the original researchers was giving a physics talk at Caltech soon after their first measurements. In those days, several Nobel physicists would sit in the front row along with Richard Feynman and other prominent physicists. About 10 minutes into the talk, Feynman walked out, much to the dismay of the speaker. About 40 minutes later, just near the end of the talk, Feynman walked back in and sat down in his seat again. The speaker then asked why he walked out and then returned. Feynman responded that he had walked out because he did not believe that the physics was correct. He explained that he had gone back to his office and worked out the problem, only to discover that the physics had been done correctly. He then returned to acknowledge the cleverness of the speaker and his colleague. Now the speaker was in dismay again, amazed that someone could have worked out the many details in so short a time!

*Brown, R. H., and Twiss, R. Q. "A New Type of Interferometer for Use in Radioastronomy."* Philosophical Magazine 45 (1954): 663.

*Silverman, M. P. A Universe of Atoms: An Atom in the Universe. New York: Springer-Verlag, 2002, pp. 102–126.*

## 240. Glauber Effect

Yes, a standard incandescent lightbulb does emit single photons, and sometimes there are photon pairs, and triplets, and so on. In the ideal chaotic photon source—a hot, incandescent wire that has physical dimensions smaller than a wavelength of the emitted light, for example—the first spontaneously emitted photon can stimulate the emission of a second photon from a nearby atom, and the two can stimulate the emission of a third photon, and so on. In principle, the photons arriving at the receptor can be single, double, triple, and so on, the actual photon state depending on how many stimulated photons were picked up before escaping the light source. The receptor receives a different energy burst with each absorption. Since the probability for stimulated emission into the same final state is proportional to the number of photons in that state already, these multiple photon processes occur quite readily.

Real light sources such as incandescent bulbs have huge physical dimensions compared to the wavelength of light. There will be numerous ideal chaotic sources along the filament wire simultaneously and randomly emitting photons toward the detector. These photons tend to arrive in bunches, with the photons within any one bunch coming from several places in the source. Very seldom does one find a steady stream of photons with nearly equal time spacing arriving from the lightbulb when one looks on the nanosecond time scale.

*Loudon, R. The Quantum Theory of Light, 3rd ed. Oxford: Oxford University Press, 2000, chap. 6.*

## 241. Bird Sounds

Some birds can emit just a fundamental frequency with no harmonics. Just how the bird eliminates the harmonics originally generated within is being investigated. The present conjecture is that a cavity resonance amplifies just the fundamental before the sound is emitted. If the fundamental frequency changes, then the cavity must change to accommodate the new fundamental in "live time."

## 242. Spouting Alligator

To eject water droplets upward, the alligator head vibrations must provide the initial energy to create nearly standing waves in the shallow water on the back of its head. As wave crests become larger, droplets of water break off and are projected high above the surface. One can simulate this effect by sliding a styrofoam cup filled with water across a finished wooden surface at about 10 centimeters per second. Water droplets will shoot upward to about 20 centimeters.

*Jargodzki, C., and F. Potter. "Spouting Water Droplets." In* Mad about Physics: Brain-twisters, Paradoxes, and Curiosities. *New York: John Wiley & Sons, 2001, p. 39.*

## 243. Hair-Raiser Function

For the HRF of a non-integer, one needs to write down a few more examples of the given integer description. Then take the logarithm of each example to discover that they all can be expressed as $\log N = n^{n-1} \log n$. By taking the exponential of both sides with the proper grouping, the final expression becomes $N = (n)^{\wedge}(n^{n-1})$—that is, $n$ to the power $(n^{n-1})$. With $HRF(x) = (x)^{\wedge}(x^{x-1})$, the HRF of non-integer values for $x$ becomes an easy calculation with the appropriate calculator, one capable of many decimal places. What is the limit as $n$ approaches zero? Complex numbers can be used, as well as irrationals such as $\pi$.

A plot of the HRF using integers shows a remarkably steep rise for even small integers; hence its name! You might want to compare its rise to an exponential function. And if all you desire is an approximate value for the inverse or for the HRF of a non-integer, the plot provides a visual image and a means to satisfy your curiosity.

However, as far as we know, the inverse HRF is awkward, and no easy calculation algorithm is known. We don't even know whether the inverse can be expressed as the limit of a series! One can determine the inverse by successive approximation to any number of decimal places with the appropriate calculator.

Of what use is the hair-raiser function? The question reminds us of two classic quotes from Michael Faraday when he was attempting to explain a discovery to the visiting prime minister. He was asked: "But, after all, what use is it?" To which Faraday replied, "Why, sir, there is the probability that you will soon be able to tax it." And when the prime minister asked of a new discovery, "What good is it?," Faraday replied, "What good is a newborn baby?"

## 244. Space Crawler

The U.S. Patent Office awarded patent 5966986 in 1999 to this propulsion device. We quote the patent abstract:

A propulsion system which is designed to be used on a payload platform such as a spacecraft, satellite, aircraft, or an ocean vessel. To operate the system electrical power is required. However, during operation the system does not require fuel or other mass be expelled into the environment to move in space. The system is designed to operate in two operational modes: in Mode I the system incrementally moves the payload platform forward with each operational cycle. In this first mode, the velocity imparted to the payload platform is not additive. In Mode II the payload platform accelerates forward a discrete increment of velocity during each operational cycle. In this second mode the increments of velocity are additive.

There is no problem with energy conservation because the onboard battery supplies the energy. The inventor Virgil Laul claims that this propulsion device when attached to spacecraft will be able to propel spacecraft out in space. We leave this problem as a final challenge. What is the physics here? Are any conservation laws violated? Will the device work in space as well as it does on the air table?

# Index

Bohr, N. *(continued)*
  double-slit, 232
  habits, 70
  quantization, 67
  religion, 65
boiling point, 134
boiling water
  altitude effects, 5
  beets in, 7
  in ice, 8
  salt added to, 7
  watched pot, 8
Boltzmann, L., 44
book rotation, 25
Bose-Einstein
  condensate, 74, 208
  Cooper pairs, 84
  He-4, 85
  statistics, 177
bosons, 85, 177
Bragg scattering, 74, 79
brain
  connections, 23
  magnetic field, 85
  power needs, 54
bread kneading, 3
Brownian motor, 50
bubble collapse, 218
bullet
  fireworks, 34
  impact, 30
Bushmen, 172
butter, measuring, 3

cadmium selenide, 59
caffeine, 7, 79
calendar
  Gregorian, 142
  Julian, 142
  lunar, 14, 143
  Mayan, 52
  rice planting, 14
calories
  from fat, 5
  human needs, 4
campfire, igniting, 134
can, pressure in, 7
car driver, 15, 42

carbon
  cycle, in stars, 245
  nuclear levels, 95
  synthesis in stars, 95
carbon-14 dating, 93
carbon dioxide
  air amount, 279
  in bread, 126
  greenhouse gas, 70, 71
  moderator, 98
  plants, 93
  in water, 9
carbonic acid, 279
Carnot cycle
  ferrofluid, 52
  photon, 61
  quantum, 61
cartoons, 30, 35
Casimir effect, 84
cellular automata, 104
centrifugal force, 111
CFC, 216
chaotic systems
  competition, 57
  hot wire, 284
  identical, 57
Chernobyl, 179
Chinese cooking, 6
chlorophyll, 135
circadian rhythm, 149
classical mechanics, 51
clocks,
  atomic, 15
  eternal, 15
  identical, 43
  light, 16
  molecular, 17
coal burning, 98
coffee, 7, 79
coherent
  light scattering, 227
  scattering, 80, 81, 176
  X-rays, 227
coin tosses
  random walk, 50
  randomness, 50
cold fusion, 98

collision
  asteroid, 36
  body cushion, 30
  bullet, 30
  molecules, 167
  spaceships, 46
  wall, in cartoons, 30
color
  cadmium, 59
  F-centers, 281
  nanophase, 59
  ruby, 120
communication
  black hole, 106
  delays, 175
  jamming, 118
  spaceships, 43, 106
computer
  atomic, 73
  DNA, 221
  java quantum, 221
  nuclear spins, 221
  quantum, 79
concrete, 239
conductivity
  electrical, 200, 220
  thermal, 130, 131, 132
conics, 21
consciousness, 86
convection, 132
Cooper pairs, 84
copper
  cladding, 175
  nanophase, 55
  oxide tunneling, 238
  X-ray laser, 73
Coriolis, G., 55
cosmic rays, 93, 124, 221
Coulomb
  barrier, 244, 245, 248
  blockade, 56
Crab Nebula, 182
creative thinkers, 85
crystal structure
  Bragg scattering, 79
  diamond, 70
  graphite, 70
  growth, 120